IN E ... ARCH

New Developments in Electrochemistry Research

Magdalena Nunez
Editor

Nova Science Publishers, Inc.
New York

For permission to use material from this book please contact us:
Telephone 631-231-7269; Fax 631-231-8175
Web Site: http://www.novapublishers.com

NOTICE TO THE READER

LIBRARY OF CONGRESS CATALOGING-IN-PUBLICATION DATA:
ISBN: 1-59454-544-8

New developments in electrochemistry research / Magdalena Nuñez (editor).
p. cm.
Includes bibliographical references and index.
ISBN 1-59454-544-8
1. Electrochemistry--Research. I. Nuñez, Magdalena.
QD555.8.N49 2004
541'.37'072--dc22
2005014622

Published by Nova Science Publishers, Inc. ✢ New York

CONTENTS

PREFACE

Electrochemistry is the branch of chemistry that deals with the chemical action of electricity and the production of electricity by chemical reactions. In a world short of energy sources yet long on energy use, electrochemistry is a critical component of the mix necessary to keep the world economies growing. Electrochemistry is involved with such important applications as batteries, fuel cells, corrosion studies, hydrogen energy conversion, bioelectricity. Research on electrolytes, cells, and electrodes is within the scope of this old but extremely dynamic field.

In: New Developments in Electrochemistry Research
Editor: Magdalena Nunez, pp. 1-53
ISBN: 1-59454-544-8

Chapter 1

ELECTROCHEMICAL PROCESSES IN TEXTILE PROCESSING

Thomas Bechtold and Heiko Brunner
Institute for Textile Chemistry and Textile Physics of the University Innsbruck;
Hoechsterstrasse 73; A-6850 Dornbirn; Austria
Phone: 0043 5572 28533; Fax: 0043 5572 28629;
Email: textilchemie@uibk.ac.at
Thomas.Bechtold@uibk.ac.at
Brunner.Heiko@dystar.com

ABSTRACT

Numerous chemical steps in textile production are based on redox reactions, for example, dyeing with vat dyes and indigo, reduction of sulphur dyes, bleach operations and waste water treatment. The introduction of electrochemical processes to replace non-regenerable chemicals by electrode processes offers various advantages, which range from better process control to savings of chemicals and water.

Generally, electrochemistry performed in textile processing is performed in aqueous solution. Direct and indirect electrode reactions have been studied and the results are given in literature. Some applications, e.g. electrochemical vat dyeing, in-situ formation of bleach chemicals, reductive decolourisation of azo-dyes have already reached technical scale operation.

Low current densities and the need for rather high cell current have led to the development of specialised cells with high active electrode area.

INTRODUCTION

Despite the knowledge of electrochemical treatments for dyes in the nineteenth century [1-10] this interesting field of research has generated a major stimulus during the last 15 years in the academic world. Nowadays electrochemical techniques are on the threshold of industrialisation.

On the one hand, there are electrochemical reductions of dyes and on the other hand there are electrochemical processes for bleaching and waste water treatments. The electrochemistry performed in textile processing in general is electrochemistry of aqueous systems. Principally direct and indirect electrochemical processes have been conceptionalised. The object of this review is an outline of this applied chemistry.

1. Direct Electrochemical Processes

Direct electrochemical processes are all-important for the cathodic reduction of sulphur- and azo-dyes. Furthermore ,some direct cathodic reductions of vat dyes and anodic oxidations, e.g. the recycling of oxidants, are described.

While azo-dyes frequently represent water-soluble systems, vat, indigoid and sulphur dyes are present as dispersed pigments. So, due to the very low probability of contact between particle and the surface of the cathode, a direct cathodic reduction is very difficult to achieve. A solution for this issue is the approach of indirect electrochemical processes using mediator-systems permitting dye reduction when beginning from the oxidised pigment. Although sulphur dyes even are pigments in their oxidised form, they represent an exception.

1.1. Cathodic Reduction of Sulphur Dyes

1.1.1. Basics of Direct Reduction and the Electrochemical Properties of Sulphur Dyes

Sulphur dyes, particularly Leuco Sulphur Black 1 and C.I. Sulphur Black 1, hold a substantial part (20-25%) of the dyestuff market for cotton textiles, especially for work-wear and denim applications, and about 70000 tons per year for example were used in 1992. [11, 12]

Generally due to the rude conditions in the synthesis of these products, they represent mixtures of oligomeric and high-molecular species that distinguish themselves by several reducible functional groups. These groups usually are disulphide groups and quinone/imine functionalities. [13,14] This is an important difference to vat and indigoid dyes which are well defined molecules that take a fixed and well defined number of electrons in the reduction step, usually two or four electrons. As indicated in Scheme 1, the disulphide groups and quinone/imine groups can be reduced to the corresponding thiolates and aminophenolates.

Thus, sulphur dyes of the quinoneimine-type can undergo several reduction reactions at different chemical sites which show similar redox potentials. [15] The investigations of Bechtold et al. and Dystar have shown, that in the case of C.I. Sulphur Black 1 six reducing equivalents and in the case of C.I. Vat Blue 43 three reducing equivalents can be received. Based on titrations of completely reduced C.I. Sulphur Black 1 using potassium hexacyanoferrate(III) and copper triethanolamine complex solutions, it could be shown that in the case of C.I. Sulphur Black 1 two electrons are needed for the reduction of the disulphidic groups and four electrons for the reduction of non-sulphidic groups, e.g. quinone/imine groups (Scheme 1). [16, 17]

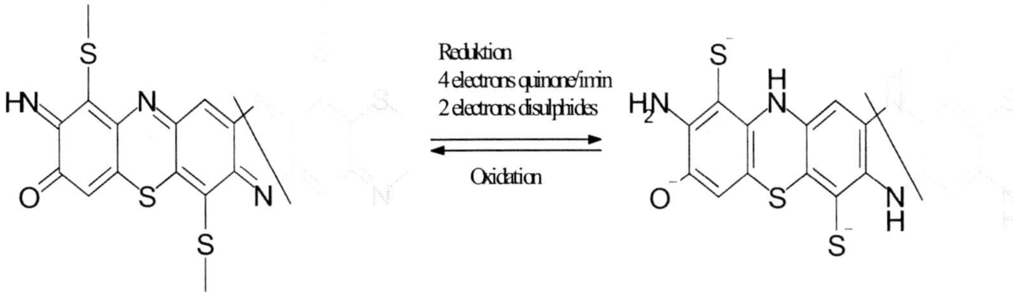

Scheme 1. Redox equilibrium of C.I. Sulphur Black 1

Experiments using the well reduced Leuco Sulphur Black 1 proved the possibility of ongoing reduction of dyestuff by addition of reducing agent (e.g. Fe(II), Sn(II) or electrochemical treatment). This indicates an important property of sulphur dyestuffs to accumulate reducing equivalents and to react similar to a redox buffer or antioxidant. [15-17]

In contrast to the vat and indigoid dyes sulphur dyes can achieve several reduction states. Unlike the vat dyes a partial oxidation or reduction of the dyestuff does not immediately result in insoluble pigments but is depending on the extent of reduction still remaining in the dyestuff. Thus in the case of C.I. Sulphur Black 1 a semi reduction (about three reducing equivalents) delivers still a homogeneous solution containing no pigments. Voltammograms measured with Leuco Sulphur Black 1 in a flow cell indicate the electrochemical activity of the dyestuff to undergo cathodic electron transfer. The possibility to achieve direct cathodic dyestuff reduction at high concentration of suspended dyestuff has been described. Batch electrolysis proves the possibility to achieve ongoing dyestuff reduction by cathodic electron transfer. In fact there can be observed an equilibrium between totally reduced Leuco Sulphur Black 1 and oxidised C.I. Sulphur Black 1. This behaviour explains the ability of such molecules to serve as a mediator and to transport reducing equivalents into the catholyte. Thus the system Leuco Sulphur Black 1/ C.I. Sulphur Black 1 represents a self mediating system and the electrolysis of sulphur dyes is an interplay of a direct electrochemical and indirect electrochemical processes (Scheme 2). [18-20]

Figure 1 shows voltammograms of Leuco Sulphur Black 1 in alkaline solution. Cathodic dyestuff reduction is achieved at cathode potentials more negative than –650 mV. In the region from –800 to –900 mV a current density of 0.25 until 0.35 mA cm^{-2} is observed.

With the insoluble oxidised C.I. Sulphur Black 1 a start up of the reduction has to be achieved at first. With increasing concentration of reducing equivalents determined in the catholyte C.I. Sulphur Black 1 gets reduced by cathodic electron transfer. Electrolysed Leuco Sulphur Black 1 and C.I. Sulphur Black 1 accumulate concentrations of reducing equivalents comparable to the state of reduction realised today in dye baths by the addition of reducing chemicals. Investigations using C.I. Vat Blue 43 prove that this described behaviour is a general feature of sulphur dyestuffs of the quinone/imine type. [17,21]

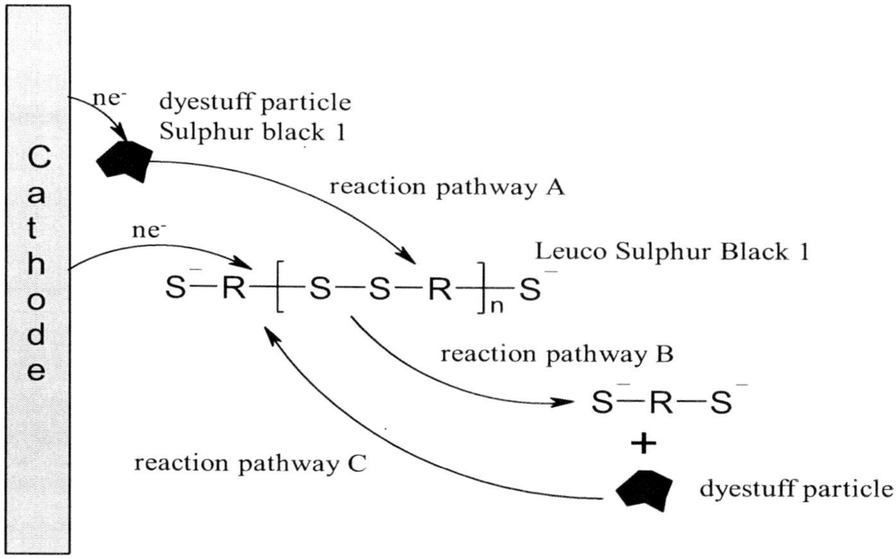

Scheme 2. Cathodic reduction of oxidised Sulphur Black 1 and soluble Leuco Sulphur Black 1

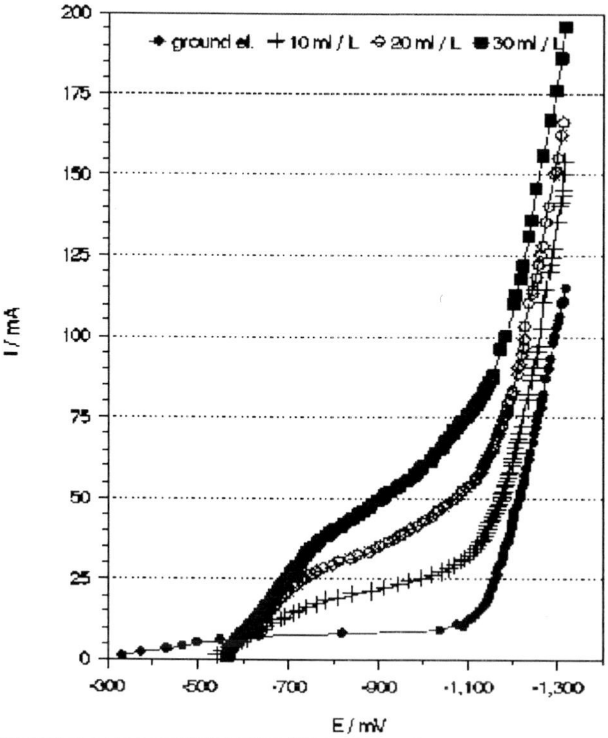

Figure 1. Voltammograms of Leuco Sulphur Black 1 (0-30mL L^{-1}) in 9.4×10^{-2} mol L^{-1} Na_2CO_3 ground electrolyte at ambient temperature

It could be shown that the properties of Leuco Sulphur Black 1 solutions concerning the colouristic data and composition depend on the reducing conditions (including the concentration sodium hydroxide and reducing agent. This aspect could also been proved for the electrolytic reduction of C.I. Sulphur Black 1 and C.I. Vat Blue 43. By using different current densities different compositions of Leuco Sulphur dyes could be obtained compared to the chemical reduction (e.g. use of sodium sulphide). In the comparison of chemical and electrochemical reductions there frequently is a difference in observed colour depth which can be explained by the different ionic strengths in the dye bath. The observed maximum in colour depth can be attributed to an optimum state of reduction/molecular weight ratio of the soluble reduced dyestuff molecules. Furthermore it can be assumed that the chemically bound molecule units forming sulphur dyes divide into smaller units by the reductive cleavage of the disulphide groups, with parallel reduction of the quinone/imine groups being proposed. The difference of compositions and colouristic properties observed between chemically and electrochemically reduced sulphur dyestuffs seems to point out that there are different or several reaction pathways for the reduction of these dyestuffs. [21]

1.1.2. Processing in Dyestuff Preparation

Sulphur dyes are obtained by the conversion of polynitrated phenoles, halobenzenes, indophenoles or indoanilines with sulphur, alkali sulphides in aqueous media. The occurring oligomeric and polymeric products consist of repeating structural elements like phenothiazone-, thiazone or phenoxazone sub-structures. [22,23] For dyeing cellulosic fibres sulphur dyes were transferred into their dyeable leuco forms by reducing agents. The low molar masses and the wide range of different molecular species are a specificum of these leuco compounds. They are soluble in aqueous alkaline media and are applicable to dyeing cellulosic fibres.

Besides the completely oxidised sulphur dyes which are immediately reduced by the addition of organic or/and inorganic reductants in the dyeing process there are pre-reduced leuco sulphur dyes as trade marks (e.g. Cassulfon™ Carbon CMR and Cassulfon™ Carbon CBR) available. These products are obtained by addition of inorganic or organic reducing agents, like sodium sulphide or glucose in alkaline media, on large scale.

In the regard of the myriads of tons as demand for sulphur dyes the chemical procedures have the disadvantage that due to the high chemical oxygen demand (COD) and the high TOC-content the demand for ecological products is still tempting.

These products are feasible by cathodic reduction. As mentioned above the leuco sulphur dye is a self-mediating system. This feature permits a production of leuco sulphur dyes starting with a small amount of reduced sulphur dyes by gradual addition of oxidised dyestuff during the ongoing electrolysis. The electrolysis can be observed by measuring the redox potential and by redox titration of samples of the catholyte. The redox potential increases as a function of transported charge from –350 mV to –800 mV when the electrolysis is started with oxidised sulphur dye. Investigations of Bechtold et al. and Dystar could show that the demanded current density depends on the concentration of the reduced dyestuff. [17]

For the cathodic reduction of sulphur dyes flow cells with plane electrodes and flow cells with three dimensional electrodes like the multi-cathode cell [17,20,21] and the precoat-layer-cell have been used.

The first electrolysis of sulphur dyes was described by Frind et al. [24]. The cathodic reductions were carried out using current densities between 5 until 50 mA cm^{-2}. Investigations of Dystar hitherto show that in the presence of high concentrations of Leuco Sulphur Black 1 current densities up to 500 mA cm^{-2} are possible. As above-mentioned the properties and molecular compositions of the chemically and electrochemically prepared leuco sulphur dyes are different. On contrary on sulphidic generated Leuco Sulphur Black 1 the electrochemically reduced Leuco Sulphur Black 1 distinguishes itself by bluer redder and brighter shades. Another very interesting aspect is that by using higher current densities leuco sulphur dyes can be obtained which distinguish themselves from chemically reduced or electrochemically reduced sulphur dyes at lower current densities.

The same colour depth of these leuco sulphur dyes in connection with lower material requirements is a feature of this technique. Table 1 and Figure 2 show a comparison between chemically and electrochemically prepared Leuco Sulphur Black 1.

Table 1. Comparison of electrochemically and chemically reduced Leuco Sulphur Black 1

Sample	Reoxidised dye dry per kg solution g kg^{-1}	L-value (brightness)	a-value (red – green)	b-value (yellow–blue)
Leuco Sulphur Black 1 (sulphidic reduced)	280	16.03	-0.33	-1.26
Leuco Sulphur Black 1 (j = 100 mA cm^{-2})	247	16.01	-0.08	-1.67
Leuco Sulphur Black 1 (j = 0.48 mA cm^{-2})	283	15.68	-0.16	-1.53

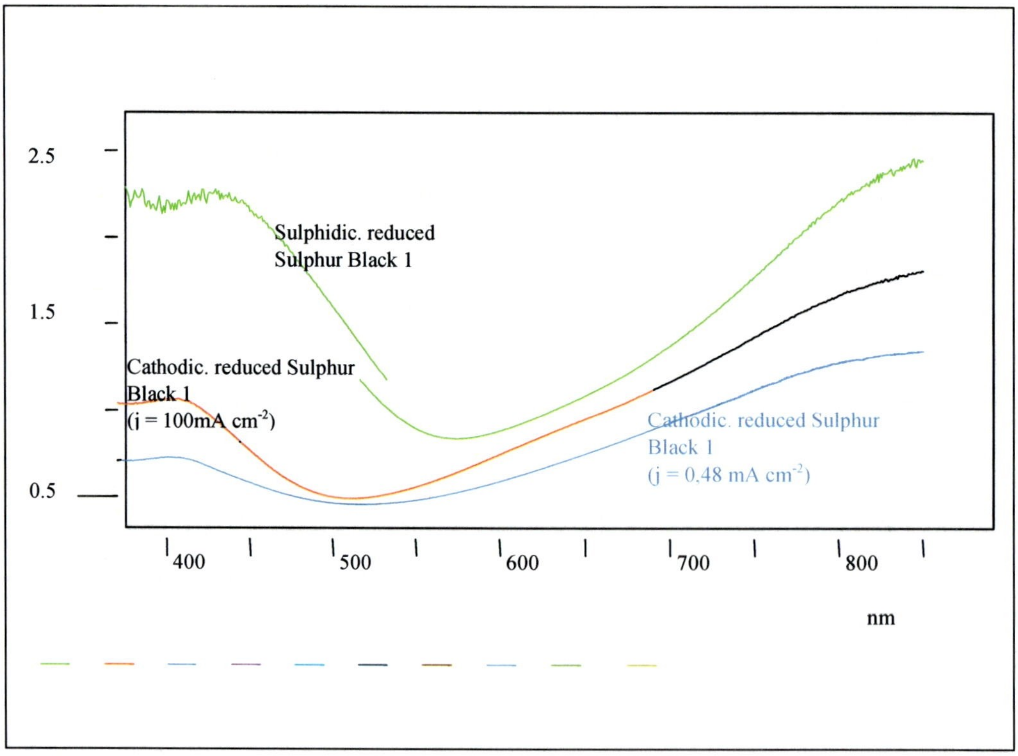

Figure 2. Comparison of UV-Vis-spectra of electrochemically reduced Sulphur Black 1

Using batch electrolysis dyestuff concentrations up to 600g/L are easily feasible. The concentration of reducing equivalents for C.I. Sulphur Black 1 is between 150Ah kg^{-1} and 534 Ah kg^{-1} and for C.I. Vat Blue 43 between 75Ah kg^{-1} and 225Ah kg^{-1}. [21]

The cathodically reduced dyestuff solutions can directly be used for dyeing processes.

2.1.2. Electrochemical Reduction in Sulphur Dyeing

Beside the use of electrochemical reduction for the production of leuco sulphur dyes on large scale the cathodic reduction can be used for the electrochemical dyeing [18,25-27]. The application of sulphur dyes in dyeing process is similar to that of dyeing with vat and indigoid dyes. The insoluble dye must be reduced by means of a suitable reducing agent to yield an alkaline soluble leuco form which has a certain affinity to the fibre. After diffusion into the fibre the insoluble and fixed oxidised form of the dye is regained by an oxidation step. Normally the reduction of vat dyes and indigo requires reducing agents that can maintain a reduction potential in the dye bath of –700 mV (vs. Ag/AgCl, 3M KCl) for indigo and potentials at least up to –900 mV for other vat dyes. Due to the lower reduction potential the reduction of sulphur dyes can be achieved with reducing agents having a reduction potential of –400 to –500 mV. Typical chemicals used in dyeing processes with sulphur dyes are as above-mentioned:

- Organic reducing agents (e.g. glucose, hydroxyacetone)
- Mixtures of dithionite with organic reducing agents and sodium sulphide.

In general each of this applications mentioned above requires the input of a certain amount of reducing agent to achieve stable reduction conditions during the dyeing process. Since all reducing agents used at this time are non-regenerable chemicals and cause great amounts of waste water containing a high chemical oxygen demand and salts, attempts were made to find techniques using regenerable systems and techniques.

The electrochemical dyeing represents a solution for these problematic ecological aspects.

According to the electrochemical properties of sulphur dyes as self-mediating systems oxidised and pre-reduced sulphur dyes can directly be used in the dyeing process. Thus a combination of electrolytic cells (e.g. multi-cathode cell) and dye-apparatus for exhaust dyeing was developed. In our investigations both pre-reduced leuco sulphur dyes and oxidised sulphur dyes as well have been used. In the case of pre-reduced leuco sulphur dyes the electrochemical reduction take on the function to compensate the re-oxidation of the sulphur dyes by ubiquitarious present oxygen during the dyeing process. The oxygen in this technical approach is reduced by the present leuco sulphur dye which is recycled by the direct self-mediated cathodic reduction. In this case the leuco sulphur dyes represent a bifunctional issue as indispensable reducing agent during the exhaust dyeing and as dyestuff. The leuco forms of sulphur dyes produced by cathodic reduction can deliver reducing capacity into a dyebath thus acting as *in-situ* reducing agents and antioxidants by themselves. Newer investigations show that the concentration of sulphur dyes need not very high to obtain good results during the dyeing [27].

The electrochemical dyeing could be used for exhaustion processes and continuous pad dyeing processes.

According to the investigations with chemical reducing agents the redox potential during the dyeing process plays an important role. As depicted in Table 2 the colour depth is dependent on the mean redox potential of the electrolyte.

Table 2. Coloristic properties of electrochemically reduced sulphur black as function of redox potential [18]

Sample	Redox-potential [mV]	L	a	b	K/S
1	-384	62.40	-2.03	-2.86	0.40
2	-393	46.41	-1.35	-4.15	1.38
3	-415	45.23	-1.31	-4.40	1.40
4	-505	32.44	-1.04	-4.50	3.33
5	-577	30.15	-0.97	-4.44	4.26

When a higher degree of reduction is achieved (measured by redox potential and titrations), the dye transports additional reducing capacity and serves as a reducing agent itself and a maximum of colour depth is reached.

During the electrochemical preparation of leuco sulphur dyes the composition of these dyestuffs can be influenced by the current density. In the same way colouristic data and shades of the dyed goods can be influenced and controlled by varying the current densities applied during the electrochemical dyeing process [21].

Considerable ecological and economic advantages are expected from a combination of low sulphide content (free-of-sulphide) leuco sulphur dyes with cathodic reduction processes in textile dyeing.

1.2. Direct Cathodic Reduction of Azoic Dyes

1.2.1. Basic Principles

More 300,000 tons of different dyes are used per year in textile dyeing operations, thus dyehouses belong to the big consumers of synthetic dyes. Dyeing almost always takes place form aqueous solution. As a result a considerable volume of coloured wasted water has to be treated in a dyehouse and legal limits have been established, which define the maximum absorbance measured at three fixed wavelength. In many European countries limits for release of textile effluents into public waste water treatment plants have been defined as spectral absorption coefficients at 436 nm (yellow) with 28 m^{-1}, at 525 nm (red) with 24 m^{-1}, and at 620 nm (blue) with 20 m^{-1}. [27]

More than 30% of the textile dyestuff consumption (110,000 t per year) belong to the group of reactive dyes which are fixed on the cellulose textile by formation of a covalent bond.

Due to the limited degree of fixation, reactive dyes for cellulose fibres form an important group of dyes which have to be considered with regard to coloured waste water. Such dyes are released during dyeing and printing operations and can be attributed to the following main sources:

- in continuous pad-batch dyeing the released residual filling of padder contains high concentrations of dyestuff [29,30]
- in exhaust dyeing hydrolysed dyes are released with the spent dyebath
- residual printing pastes contain a high amount of reactive dyes [31]
- washing processes following to the dyestuff application lead to the formation of a huge amount of diluted coloured waste water.

When modern formulations of dyes are used the toxicity of the dyes in the wasted water is a minor problem, however the limited biodegradation of the dyes in communal waste water treatment plants has led to definition of limits for absorbance of coloured wastes cited above [28].

The treatment of mixed waste water collected from various sources in a dye house has been the usual technique for waste water treatment for a long time [32].

Nowadays the strategy changed towards identification and separation of waste water streams which contain a distinct content of chemicals and thus need further treatment. Many reactive dye structures contain azo groups and reductive cleavage of the azo chromogene leads to a decrease in specific absorbance of the dye [33,34].

The azo group can be reduced both chemically and by cathodic electron transfer, in the literature studies were performed to describe the behaviour of selected model dyes and commercially significant reactive dyes during cathodic treatment. In particular the electrochemical properties of azo-groups have been studied by numerous research groups.

Eriksson and Nyholm studied the electrochemical behaviour of azo-salicylic acids on a glassy carbon electrode and showed that depending on the structure of the molecule is reduced by two electron reduction to the hydrazo form or in an overall four electron reduction to the corresponding amino form [35,36].

Indirect cathodic reduction of Acid Yellow 9 using anthraquinone-2-sulphonate also yielded the corresponding hydrazo form, which could not be reduced further in the case of Acid Yellow 9 [37].

Electrochemical properties of a series of azobenzene dyes also were studied by Podsiadly and other groups [38-42]. A detailed analysis of the voltammetry of diazo-reactive dyes is given in the literature. As a general finding the cathodic reduction of diazo-dyes totally consumes eight electrons per dyestuff however proceeds via a first four electron reduction to the bis hydrazo form which then either disproportionate or rearranges with cleavage of the molecule and formation amino compounds [43,44].

The reduction of the azo group can proceed according to reactions (1) and (2). The use of the Fe(II)-triethanolamine complex as a very powerful reducing agent permits the evaluation of working conditions useful for the electrochemical dyestuff reduction [45]. Using this reducing agent a plateau in the redox potential can be observed in the redox titration curve. This plateau can be attributed to the reduction of the azo group.

(1) $\text{Ar-N=N-Ar'} + 2e^- + 2H^+ \rightleftharpoons \text{Ar-NH-NH-Ar'}$

(2) $\text{Ar-NH-NH-Ar'} + 2e^- + 2H^+ \rightleftharpoons \text{Ar-NH}_2 + \text{Ar'-NH}_2$

When the actual dyestuff content and the chemical structure of the dyestuff is known the titration data permit to distinguish whether full dyestuff reduction is achieved according reaction (1) and (2) or if the reduction stops with formation of the hydrazo group (1).

C.I. Acid Red 27 and C.I. Acid Yellow 9 have been investigated as model systems of azo dyes with known purity. Typical representatives of commercial reactive dyes with azo groups were investigated on condition that at least the chemical structure has already been given in the literature. With use of a divided cell as electrolyser, decolourisation of model waste water can be achieved without addition of chemicals, which is of interest with regard to minimising chemical consumption in waste water treatment and to get a cost effective decolourisation process. The technique is of particular interest for waste water decolourisation because in many cases the addition of chemicals, flocculation auxiliaries, leads to a formation of considerable amounts of precipitate which has to be separated and disposed with additional costs [44,46].

1.2.2. Decolourisation of Waste Water from Reactive Dyeing

A detailed analysis of the sources for coloured waste water released by a dyehouse indicated that 15 - 30 % of the applied dyestuff are released into the waste water. The main part corresponding to 50 - 80 % of this amount result from washing/rinsing steps and thus the major part of the dyestuff found in diluted solutions with average concentrations of 1 - 2 g L^{-1}. The part of dyes decolourised by collecting padder fillings thus represents only a minor part of the total amount of released dyestuff [32].

To achieve efficient treatment of such wastes a removal and reconcentration of the dyestuff is performed by means of Nanofiltration (NF) which permits a removal and reconcentration of the dyestuff to concentrates with approximately 5 g L^{-1} dyestuff [47-51]. Parallel the volume of the dyestuff containing concentrate decreases to approx. 10 % of the initial volume. Such treatment leads to considerable advantages with regard to water recycling however for the treatment of the concentrates research is needed to establish suited processes.

Due to the reconcentration and the volume reduction cathodic dyestuff destruction could be an interesting procedure for the treatment of NF-concentrates.

Another source for dyestuff containing wastes is found in textile printing operations [52]. The residual printing pastes and the washing water from cleaning of screens and printing equipment contain high concentrations of dyes but additional problems result from the presence of thickeners to increase viscosity and auxiliaries [53-56].

The cathodic reduction of azo-group containing waste water from reactive dyeing has been proved to be a suited process for decolourisation of concentrated wastes from pad-batch reactive dyeing, particularly for decolourisation of residual fillings of padders which normally are withdrawn at the end of the impregnation process. In the literature basic investigations and successful scale-up trials have been described [29,30,46,57].

In the scientific and technical studies cathodic dyestuff reduction was applied to the decolourisation of:

- dyestuff containing baths from reactive dyeing (exhaust dyeing) and washing operations (continuous and exhaust dyeing) which were reconcentrated by Nanofiltration. Such solutions contain salt, auxiliaries and comparable low amount of alkali [56].

- printing pastes from reactive dye printing, collected from technical textile printing operations (drain of screens, vessels and containers for pastes). These mixtures contain thickener, auxiliaries and weak oxidising agents (e.g. m-Nitro-benzene-sulphonic acid sodium salt) [53].

At present a full scale electrolyser for the cathodic treatment of such wastes has started operation in a textile dyehouse (cell current approx. 250 A).

The reduction of azo-groups in synthetic dyes successfully could be developed to full scale operation. Separated cells with cation exchange membranes as separators have been built and tested in various scale. The cell current is limited by the concentration of the dyes present in the catholyte thus an efficient use of the technique is possible when relative concentrated dyestuff solutions are available. In addition cells with high cathode surface are required to keep installation costs reasonable.

The need for high cathode surface led to the development of multi cathode electrolysers, which combine high cathode area with relative low membrane and anode area.

Typical data measured at different cell types decolourising either model solutions or technical waste water are given in Table 3. Figure 3 gives a general reaction scheme for the cathodic decolourisation of azo dyes. Figure 4 shows a scheme of a multi cathode electrolyser for waste water decolourisation.

Tests in technical scale to decolourise concentrated reactive dye solutions containing 10 – 50 g L^{-3} reactive dye showed that a capacity of 1 m^3 / day of the concentrated solution (50 kg of dyestuff) could be decolourised with a multi-cathode electrolyser at a cell current of 150 A (Figure 4 and 5). The capacity of 1 m^3 per day corresponds to the full volume flow of such concentrates released from a medium size dyehouse. Average decolourisation of 80.0 % at 445 nm, 85.5 % at 520 nm, 69.4 % at 620 nm and 78.3 % at 660 nm was realised. The lower degree of decolourisation at 620 nm is due to the presence of stable blue-green dyestuffs which contain phthalocyanine or anthraquinoide groups in their chromophoric system.

Energy consumption of the pilot cell was determined near 20 kWh m^{-3}. A further lowering of the electrical energy consumption is expected to result from an optimisation of the cell geometry, e.g. by decreasing the distances between cathodes anode. Experiments at elevated temperatures were not performed because even a low increase in temperature e.g. from 20°C to 40 °C will consume 23 kWh m^{-3} and thereby will double the overall consumption of energy [44,46,57]. At a cell current of 130 A the consumption of NaOH in the anolyte was 0.25 L of 50 % NaOH solution per hour.

Based on tests in full scale a multi cathode electrolyser for decolourisation of both concentrated solutions from pad-batch reactive dyeing and concentrates from NF-treatment was constructed and installed in a dyehouse. The capacity of the 250 A cell is 2 – 3 m^3 per day dyestuff solution from pad batch dyeing or 3 – 5 m^3 per day dyestuff concentrates obtained from nanofiltration treatment of the diluted effluents. At fixed cell current overall decolourisation mainly is dependent on the time of electrolysis. In full scale operation an energy consumption between 2 – 8 kWh m^{-3} of waste water was determined during 8 h batch electrolysis. Decolourisation was between 60 and 80 % of the initial absorbance. Optimum cell performance was observed when mixtures of nanofiltration concentrates and residual padder solutions were treated because of the positive effect of the alkali brought into the mixture by the padder solution. A general operation scheme is shown in Figure 6.

Table 3. Cathodic decolourisation of azo dyes - Working conditions and cell data

Cell type	Solution	Cell Voltage	Cell current	Cathode		Current density	Anode		Capacity	Energy consumption	% decolouris.
					area			area			
		V	A		m^2	A m^{-2}		m^2	l $time^{-1}$	kWh m^{-3}	%
Lab. scale - Flow cell, plane electrodes [46]	0.5 g L^{-1} Acid Red 27	3.5	0.02	stainless steel	0.01	2	stainless steel	0.01	0.13		
Lab. scale - Multi cathode electrolyser [44]	1-10 g L^{-1} reactive dye	5 – 7	0.9 - 3	3 isolated cathodes stainless steel mesh	0.43	2.1 – 7.0	stainless steel	0.01	0.25 l/h	10 – 80	80
Lab. scale - Multi cathode electrolyser [46]	0.5 g L^{-1} reactive dye	5.2-7.5	5	10 isolated cathodes stainless steel mesh	1.9	2.6	Pt-mixed oxide coated coated Ti-electrode	0.04	2 l h^{-1}	3.7 – 7.8	50
Pilot scale - Multi cathode electrolyser [30]	5-15 g L^{-1} reactive dye	5 - 10	120-150	2x10 isolated cathodes stainless steel mesh	17.2	7.0 - 8.7	Pt-mixed oxide coated coated Ti-electrode	0.39	50 l h^{-1}	20	80
Technical scale - Multi cathode electrolyser [57]	padder filling, nanofiltration concentrates 5 – 50 g L^{-1} reactive dye	3.1 – 4.1	80 - 250	2x2x5 isolated cathodes stainless steel mesh			Expanded mesh steel anodes		200 l h^{-1}	2 – 8	70-80

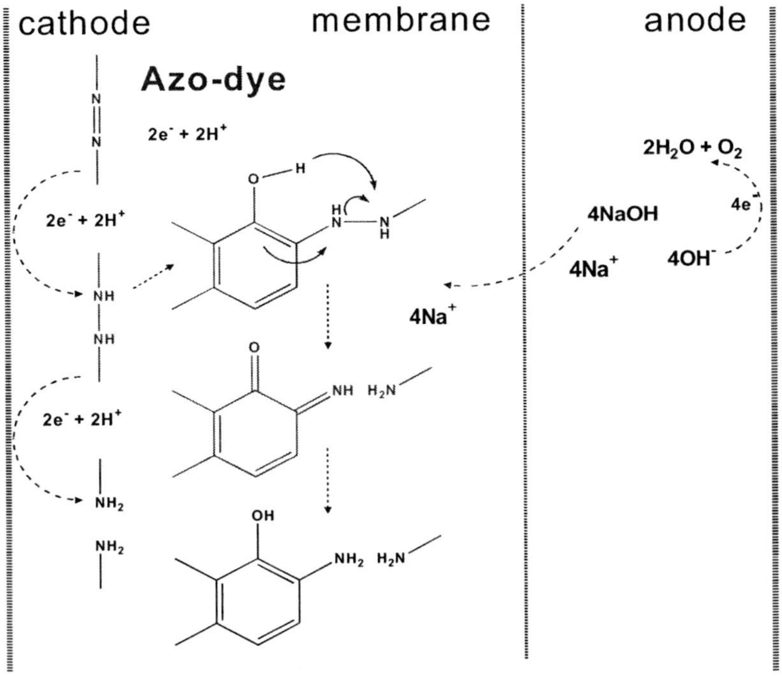

Figure 3. Reaction scheme for the cathodic decolourisation of azo dyes.

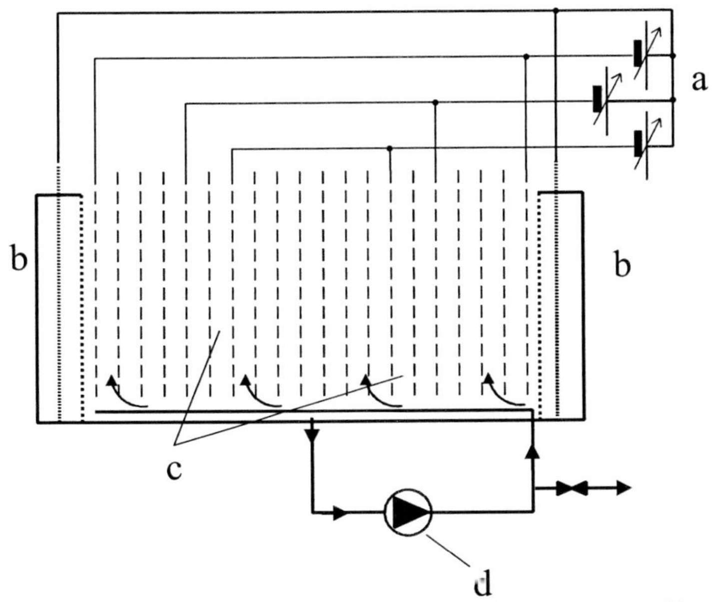

Figure 4. Batch electrolyser - technical scale. a power supply, b anode compartment with cation exchange membrane, c cathode units, d catholyte circulation.

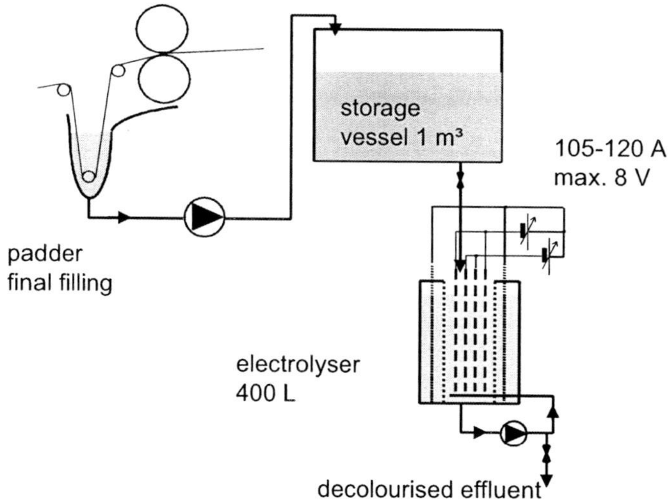

Figure 5. Scheme of a installation for cathodic decolourisation of residual padder fillings (approx. 50 g L^{-1} reactive dye, capacity 1 m^3 per day).

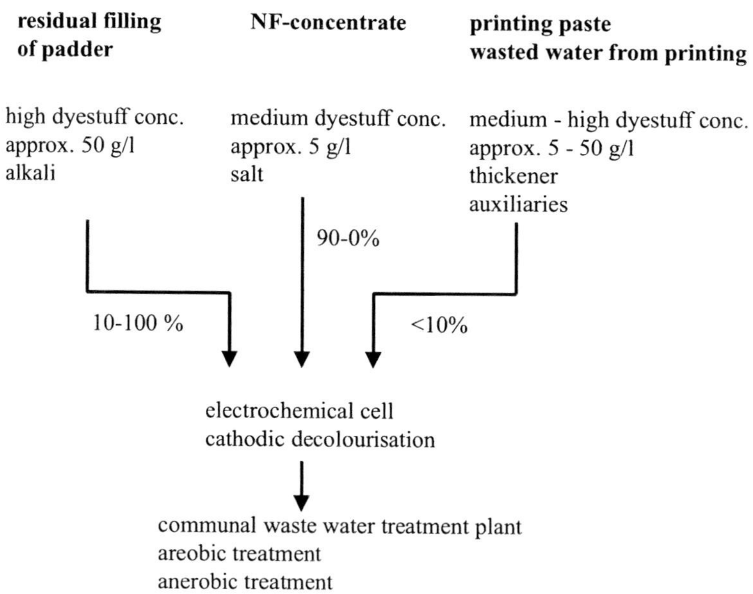

Figure 6. Scheme for the decolourisation of coloured wastes form textile dyehouses.

The addition of mediator systems e.g. based on iron-triethanolamine/D-gluconate complexes increases decolourisation rate at fixed conditions which could be used to lower cathode area and to achieve higher extent of decolourisation. However even when low amounts of chemicals are used any addition of mediator will increase chemical load in the waste water.

Figure 7. Top view of a 250 A multi-cathode electrolyser for cathodic decolourisation of residual padder fillings.

Figure 7 shows the top view of a 250 A multi-cathode electrolyser installed in dyehouse for cathodic reduction of azoic-group containing reactive dyes.

In this cell a volume of 400 L concentrated dyestuff solution was electrolysed with 120 – 130 A cell current. The absorbance of the catholyte was monitored photometrically (path length 0.1 mm). In Figure 8 the decolourisation in terms of reduction of absorbance as function of electrolysis time is shown.

The use of divided cells is important to separate the anodic process from the cathodic waste water treatment because in many cases wasted water from reactive dyeing contains chlorides in considerable amounts thus anodic formation of chlorine could cause critical acceptable concentrations of AOX in the waste water.

NaOH solution in the concentration range of 0.5 – 1 mol L^{-1} NaOH is used as anolyte thus as by product of electrolysis NaOH is formed in the catholyte.

1.2.4. Decolourisation by Anodic Processes

Another group of electrochemical processes discussed in the field of dyestuff removal in wasted water bases on the formation of precipitated metal hydroxides by electrochemical dissolution of iron or aluminium anodes. The precipitate formed strongly adsorbs dyes present in the solution, in some cases dyestuff reduction and oxidation processes are described to occur in the electrolyser in parallel to sludge formation. The processes result in formation of a considerable amount of sludge to be disposed thus additional costs have to be considered [58--62].

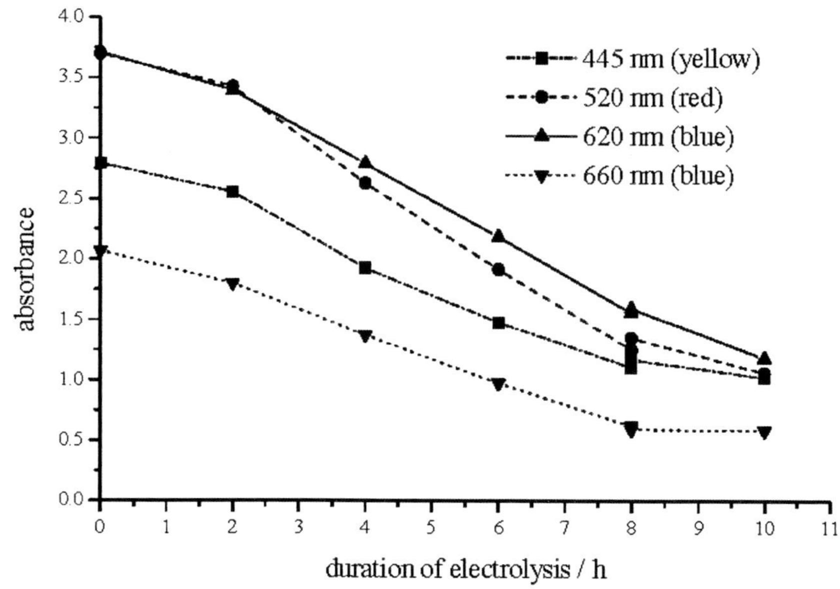

Figure 8. Decolourisation of 400 L technical dyestuff solution from pad-batch dyeing in a multi-cathode electrolyser at 120-130 A cell current.

Electrochemical oxidation processes also have been described to be applicable for dyestuff destruction, however clear distinction has to be made between indirect electrolysis by formation of oxidants like chlorine/hypochlorite and direct anodic oxidation of dissolved organic matter. The anodic formation of hypochlorite to decolourise wasted water permits excellent decolourisation rates however possible formation of toxic products e.g. AOX (adsorbable halogenated organic compounds) has to be considered thoroughly [63-64].

2. Direct Reduction of Vat Dyes

2.1. Basic Principles and New Developments

Vat and indigoid dyes are extensively used in the coloration of cellulosic fibres. The great advantage of this class of dyestuffs is their all-round fastness, particularly to light, washing and chlorine bleaching and they represent in the regard of quality a real alternative to reactive dyes [12]. As mentioned above vat dyes require a complicated and sophisticated dyeing procedure because of their insolubility in aqueous media. The typical approach consists of a reduction step using a chemical reducing agent to obtain the water–soluble leuco form of the dye and after diffusion into the fibre a re-oxidation step to become pigmented. In analogy to the sulphur dyes ecologically contaminated waste water is obtained by this procedure including a high chemical oxygen demand. Therefore many attempts have been made to replace the environmentally unfavourable chemical reductants like sodium dithionite by ecologically more attractive alternatives. The electrochemical treatment represents one of the most efficient methods. Although Grotthuss [1] described the first electrochemical reduction of indigo in 1807 and Goppelsroeder [2-8], Wartha [9], Mullerus [10] and Nevyas [65] extensively studied the cathodic reduction in the end of the nineteenth century the electrochemical vatting has slept a long torpor. Unfortunately the current yields were always less than 20% and thus an industrial application was prohibitive however new promising efforts and studies were undertaken in the last decade.

In their oxidised form indigo and vat dyes have a pigmented character and thus they are not soluble in water or aqueous solutions. Thus the direct cathodic reduction is very difficult to achieve because of the very low probability of a contact between dye particle and the surface of the cathode [15]. In contrast to the sulphur dye there is no interplay between a direct cathodic and an indirect electrochemical process. Due to the well defined molecular structures vat dyes have well defined reduction states. Although Rys et al. describe a radical mechanism and an interaction between the dye and the leuco dye (*via* a radical anion) [66-68] no significant self mediating effect like in the case of sulphur dyes could be observed in the same extension. Due to the very low probability of a contact between the oxidised vat dyes particle and the surface of the cathode the used current densities have to be very low to avoid side reactions (e.g. water electrolysis) and to obtain good current yields. Unfortunately this approach results in bad or moderate space time yields. Thus the economy of these processes is exactly unfruitful for the reduction of vat dyes and indigo.

Concerning the direct reduction one can determine two different technical approaches:

- the direct cathodic reduction and
- the electrocatalytic hydrogenation.

The direct cathodic reduction is one of the first approaches of electrochemical treatments of dyes at all. As above-mentioned several approaches have been made. Motivated by the results of Charmaut [69,70] Rys and Roessler have investigated the use of graphite as a high-surface-area material for the reduction of vat dyes and indigo. In dependence on the good efforts using a three-dimensional electrode in form of a precoat-layer cell by BASF (Figure 9) [71,72] they have used an electrochemical fixed- or fluidised-bed reactor [73-75].

Using graphite no chemisorption and electrocatalytical hydrogenation is possible. Unfortunately the reduction rates were very low. A solution for this problem was found by the modification of the graphite. By oxidative treatment (e.g. soaking with hydrogen peroxide pre-anodisation) the graphite could be activated due to the formation of quinone and hydroquinone functionalities that are described as mediator systems for the reduction of vat dyes [73]. In correspondence to this results the covalent bounding of quinoid molecule like 1,8-dihydroxyanthraquinone or 5-amino-acenaphthenequinone, could enhance the reduction rates [73-75]. Due to the oxidation of the graphite the obtained carboxy-, hydroxyl-functionalities could be functionalised and thus used for the immobilisation of suitable electron transfer mediators. Thus several immobilisation concepts are thinkable and a suitable one depends on the graphite and/or the mediator molecules.

As depicted in Scheme 3 several anthraquinone-derivatives have been used in the reduction of vat dyes and indigo.

In this way the method represents a hermaphrodite between a pure surface-based and a mediator catalysed reduction. Regarding the very alkaline conditions (pH 12-13 and 80°C) in vat dyeing a saponification of these ester- and amide-bounded mediators can not be ruled out and thus the existence of free and well investigated mediator molecules is possible during the process.

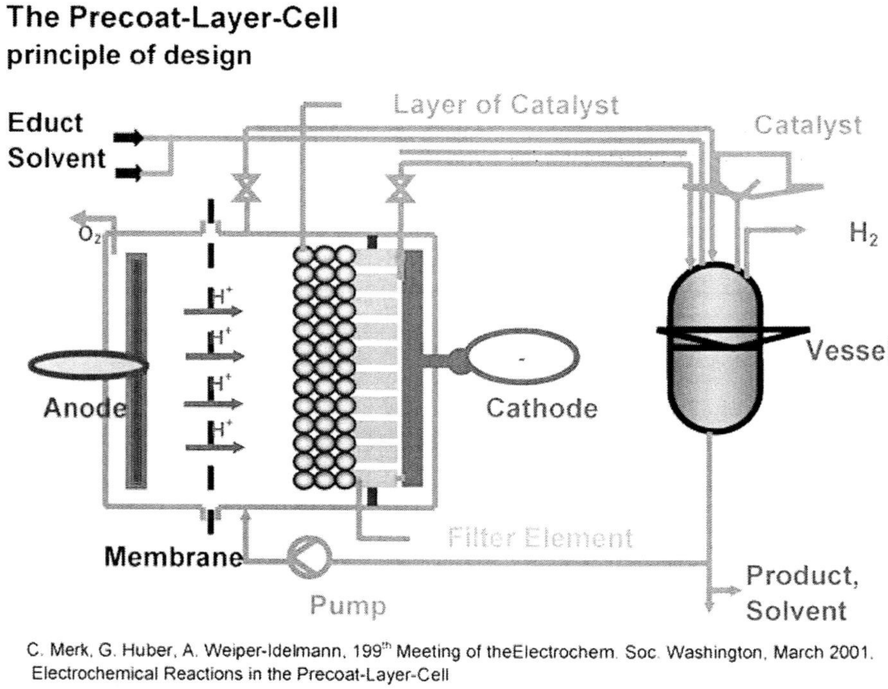

Figure 9. Precoat-layer-cell

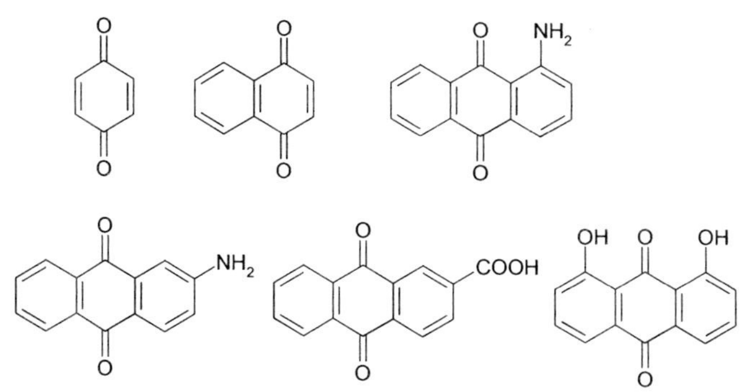

Scheme 3. Used immobilised mediators

Anyway using this approach the reduction of indigo and vat dyes was possible with current yields up to 86% in the presence of low current densities.

In accordance with the results of Goppelsroeder, Wartha, Mullerus and Nevyas new investigations in the DyStar Laboratories show the general possibility of direct cathodic reductions [17]. Using a new developed electrolytic cell equipped with three-dimensional multiple expanded stainless steel electrodes the reduction of indigo was generally possible using low current densities up to 80%. Unfortunately the enhancement of the current densities to improve the space-time-yields leads to bad and insufficient current yields. Thus a direct cathodic reduction of vat dyes rules out in the regard of industrialisation (Figure 10).

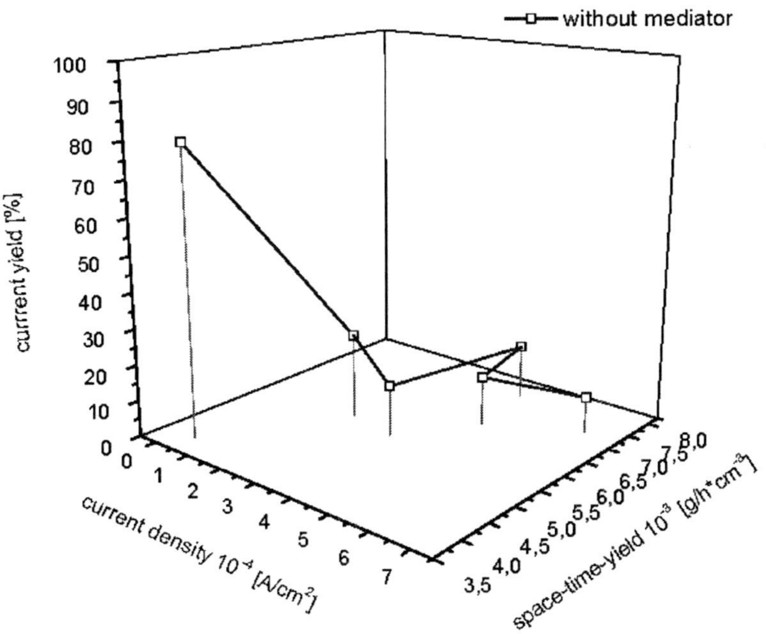

Figure 10. Direct cathodic reduction of indigo – subordination of current yield and space time yield to current density.

2.1. Electrocatalytic Hydrogenation

The electrocatalytic hydrogenation which has found a broad scope in electrochemistry was applied to the reduction of vat and sulphur in the last nineties of the recent century and it proves to be a promising and attractive alternative [71,72,76,77]. In contrast to the described direct and indirect processes the electrochemical hydrogenation consists of a process in which the electrochemically *in-situ-* generated hydrogen in *statu nascendi* directly reacts with the organic substrates (i.e. vat dyes). The hydrogenation-step is in competition with the hydrogen evolution reaction and thus the efficiency of this process depends on this competition. Based on the fact that the hydrogen is available in *statu nascendi* the kinetic barrier of splitting the hydrogen molecules and chemisorption of these atoms is completely bypassed and elevated pressures and temperatures can be avoided.

Rys and co-workers have studied the industrial feasibility of this approach in a divided flow cell using Raney-nickel electrodes in the reduction of indigo. Raney-nickel brought the availability, costs and stability in alkaline media into accord. Despite of the possibility of the reduction of several vat dyes this method failed because of the low current efficiencies of about 13% in the case of indigo. Since a huge electrode surface of more of several hundreds of square meters would be necessary to obtain an industrial relevant reduction rates and efficiencies for stock solutions, this method contradicts any reasonable industrial use. Alternatives like platinum-black show only insufficient results due to their low stability which ruled out an industrial application [76,77].

Another efficient approach concerning the reduction of indigo and vat dyes was reached by the use of the precoat-layer-cell which was developed by BASF (Figure 9) [71,72].

The electrolytic cell is based on the filtration principle. In this approach the indigo is brought into contact with the cathode, where the cathode is formed by a support of an electrically conductive material (i.e. filter fabrics) and an electrocatalytical layer, like palladium on carbon, Raney-nickel, Raney-iron etc., formed on a filter in situ by precoat filtration.

Using palladium on carbon current efficiencies up to 80% could be obtained in the case of indigo.

Despite of good results in the electroorganic chemistry the industrial use of this electrolytical cell has not established yet [78].

3. Electrochemical Formation of Oxidants

3.1. In-Situ Formation of Oxidants

Numerous electrochemical processes for production of oxidants are given in literature however the major part is not used for in-situ formation of an oxidant to serve as bleach chemical.

The on-site and in-situ formation of hypochlorite has been studied extensively, mainly the investigations were directed towards formation of hypochlorite containing solutions and later use in bleach processes [79,80]. Anodic formation of hypochlorite also has been studied for textile wash/bleach processes [81,82].

Considerable research for the application of anodic formation of oxidants has been done for pulp bleach processes [83-85]. Organic mediator systems e.g. based on violuric acid have been used for anodic bleach processes on textiles and for electrochemical delignification [86,87].

Oloman and co-workers studied the application of manganese or iron complexes as mediators for pulp bleach processes [88].

Another group of oxidants formed electrochemically bases on cathodic formation of hydrogen peroxide by reduction of oxygen [89].

3.2. Electrochemical Bleach Processes

Recently use of anodic processes for oxidative bleach processes have been studied more intensively. Processes for electrochemical production of bleach chemicals have been studied in the literature for delignification of pulp and for textile bleach processes.[86,90-96] The electrochemical formation of hydrogen-peroxide on-site has been investigated [97]. The use of electrochemical methods to form organic oxidizing agents for denim bleach has been described in the literature [86]. Limited scientific information is available about the electrochemical in-situ formation of hypo-halogenites for bleach operations [79,93,98,99] The in-house preparation of NaOCl also has been described in the literature, however the efficiency of the bleach process has been described as rather low [95].

An interesting application of electrochemical bleach could be the oxidative decolourisation of denim fabric in jeans production.

In-site formation of hypochlorite solution for denim bleach offers several advantages to the applicant:

- defined conditions of the bleach leading to better reproducibility and process cost saving
- no losses of activity of hypochlorite during storage
- better defined conditions of bleach process
- improved reproducibility of bleach step
- regeneration of bleach bath is possible.

Current efficiency of the electrolysis is depending on the duration of the electrolysis respectively the concentration of oxidant formed. Laboratory experiments with anodic current density of 60 – 500 A m^{-2} showed only minor influence of current density on current efficiency. Up to a concentration of "active chlorine" c(Cl) = 0.1 mol L^{-1} current efficiency is determined with 60 to 80 %.

For denim bleach processes "active chlorine" concentrations of c(Cl) = 0.12 mol L^{-1} corresponding to 30 ml L^{-1} commercial NaOCl solution are used. A simple and robust unipolar cell without mechanically sensitive components, designed cell for a current of 1000 – 1200 A permits formation of hypochlorite solutions of the required concentration level. Anodic current density was adjusted near 400 A m^{-2}.

A significant increase of the bleach process could be observed in presence of low concentrations of added bromide [94]. Due to the higher reaction rate between hypobromite and indigo the addition of bromide results in a distinct intensification of the bleach effect.

A scheme of the reactions occurring can be proposed according to equations (3) to (5).

$Cl^- + H_2O \rightleftarrows OCl^- + 2H^+ + 2e^-$ (3) anode reaction

$OCl^- + Br^- \rightleftarrows Cl^- + OBr^-$ (4) homog. reaction

$OBr^- + \text{indigo} \rightleftarrows Br^- + \text{isatin}$ (5) heterog. bleach

Accelerated bleach reaction in presence of OBr^- intensifies decolourisation of indigo, despite the fact that the molar concentration of Br^- 8.4 10^{-4} mol L^{-1} can be less than 0.5 mole % of the NaCl concentration of 0.171 mol L^{-1} Cl^- in the electrolyte. The current density of the anodic process is determined by the concentration of chloride ions, but the bleach effect in solution is intensified by the addition of bromide.

The current efficiency of the anodic process is defined by the heterogeneous charge transfer reaction between anode and chloride ions (eqn. 3). However the formation of the more active hypobromite is the follow-up reaction in solution (eqn. 4). Due to homogenous conditions of reaction 4 low concentrations of bromide are sufficient for a distinct intensification of bleach effect.

The more intensive bleach effect indicates a "catalytic" action of the Br^- / OBr^- system which continuously is regenerated according to equation 4 and 5.

The increase in bleach activity also causes differences in the shade of the bleached goods. The colour of the bleached denim tends towards more yellow/greenish shades when bromide is present in the anolyte. Thus the selection of an appropriate bleach reagent also will be determined by the required final shade of the bleach process.

3.3. Technical Concepts

Two basic technical concepts for in-situ formation hypochlorite in denim bleach are in consideration at present [96].

- operation as a stand-by cell
- online operation

When the cell is operated as stand-by unit the bleaching liquor is prepared before the wash/bleach process either as the total bleach liquor or as concentrated hypochlorite solution. This concept permits the use of smaller cells operating with lower cell current because longer duration of electrolysis for production of the required amount of oxidant is permitted. In addition a number of washing machines can be supplied by a single installation.

Online operation requires construction of a compact flow-through cell coupled to drum washer. The integration of the cell permits direct steering of the bleach process by selection of cell current and total transported charge, however the cell capacity has to be sufficient to produce the required amount of oxidant within the planned bleach time.

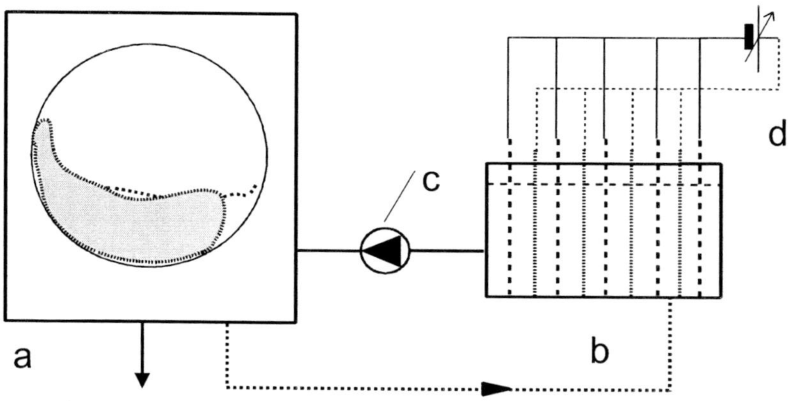

Figure 11. Scheme of the coupling between electrolyser and drum washing machine: a drum washing machine, b electrolyser, c filling pump, d power supply.

For full scale tests a undivided cell manufactured by DE NORA Elektrochemie (HKE-4-H, DE NORA Deutschland, Rodenbach, Germany) was used. The cell was built as unipolar cell equipped with parallel electrodes (5 cathodes and 4 anodes). A Tonello G1 150 drum washing machine with a load of 30 kg textiles and a filling of 300 L of bleach bath was coupled to the cell (Tonello, Sarcedo, Italy) (Figure 11).

The size of the electrolyser was defined by the technical demand to produce the required amount of chlorine bleach within the time set for a bleach process in the drum washer. For an estimation of the capacity of the electrolyser a concentration of 56.4 mM NaOCl, equivalent to 30 ml L^{-1} commercial 1.88 M NaOCl solution, should be produced in a volume of 200 L within 60 min. Assuming a current efficiency of 50 % a cell current of 1210 A was calculated as cell current required to produce 22.6 moles of oxidant within 60 min.

In full technical scale tests NaCl solutions with a concentration of 0.236 - 0.944 M NaCl (13.8 g L^{-1} to 55.2 g L^{-1}) were electrolysed to obtain hypochlorite solutions of c(Cl) = 0.10 – 0.12 M („active chlorine“) at current efficiency 58 – 70 %.

Energy consumption per kg of "active chlorine" formed was determined with 5.7 – 8.6 kWh kg^{-1} [100].

By coupling the electrolyser to a commercial drum washing machine with a load of 30 kg of denim garment the electrochemically prepared oxidant solution was used for bleach processes. In comparison to commercial hypochlorite solution electrochemically prepared bleach baths showed more intensive bleach effect.

Figure 11 shows a scheme of an installation to couple the hypochlorite electrolyser with an drum washer for denim garment.

Table 4 gives an overview about relevant process and cell data used for the in-situ production of hypochlorite.

Table 4. Geometric data of the laboratory scale electrolyser and general experimental conditions

	Experimental parameter	Laboratory scale	Technical scale
Anode	material	1 electrode Pt-mixed oxide coated titan expanded mesh	4 electrodes 0.4 m² Pt-mixed oxide coated titan expanded mesh
	geometric area (both sides considered)	400 cm²	4 x 0.4 m² = 1.6 m² 2 x 4 x 0.4 m² = 3.2 m²
Cell current	total current	5 – 20 A	250 – 1200 A
	anodic current density	12.5 – 50 mA cm^{-2} 125 – 500 A m^{-2}	7.8 – 37.5 mA m^{-2} 78 – 375 A m^{-2}
Cathode	material	1 electrode stainless steel fabric	5 electrodes 0.4 m² stainless steel expanded mesh
	geometric area	360 cm²	5 x 0.4 m² = 2.0 m²
	cathodic current density (both sides considered)	140 – 560 A m^{-2}	125 – 600 A m^{-2} 62.5 – 375 A m^{-2}
Electrolyte	volume	11 L	160 L

In the technical scale tests a cell voltage of 4.5 – 6.8 V was required for a cell current of 1000 A. On basis of an average current efficiency of 60% production of 1 kg of chlorine requires a charge flow of 1260 Ah respectively consumes 5.67 – 8.57 kWh kg^{-1}. This corresponds to a volume of 7 L commercial NaOCl solution (150 g L^{-1} NaOCl).

3.4. Oxidative Waste Water Treatment

Anodic hypochlorit formation for destruction of dyes has been studied by several groups [63,101]. The anodic decolourisation in presence of chlorides results in excellent effects however the risk of formation of AOX (adsorbable halogenated organic compounds) in the wastewater requires critical consideration of the process conditions.

A promising technique intensively studied recently is based on the use of boron-doped diamond electrodes as anodes for oxidation of textile dyes. Particularly the oxidative anodic decolourisation of organic compounds including soluble dyes has been studied in detail [102-108]. In many cases sulphate is present as ground electrolyte and the formed peroxodisulphate is used for oxidative destruction of the dye in a following treatment step [107]. While for scientific studies anodic processes can be performed at low concentrations of chloride in many dyehouses the practical application of textile dyes particularly reactive dyes is coupled to the addition of considerable amounts of NaCl.

The oxidative decolourisation of various dyes including indigo has been studied on Pt-electrodes [109]. On boron-doped diamond electrodes (BDD) the anodic oxidation of organic compounds proceeds via the formation of hydroxide radicals, thus the electrochemical incineration also can be used for dyestuff oxidation [110].

4. Indirect Electrochemical Reduction

4.1. Redox Couples – Chemical Properties

Indirect electrolysis of vat dyes, indigo and sulphur dyes is of interest to achieve higher current densities for the reduction of the dispersed dyestuff pigments. These dyes are separated into three classes of dyes because of their characteristics in application and final product. Vat dyes yield coloured fibres with excellent colour fastness and the position of indigo in the production of denim is almost unique. Despite the fact that indigo is a vat dye its main application for the unique denim market (jeans) requires special techniques and conditions during its application. Sulphur dyes are economical and show good fastness for most purposes [111-115].

While the final properties of the products differ the application of vats, indigo and sulphur dyes follows a common general scheme demonstrated in Figure 12. All dyes required the use of reducing agents during the application step to obtain the alkali soluble reduced form which exhibits considerable affinity to the cellulose fibre, required for the dyeing process. Important reducing agents in technical use today mainly are: Sodium dithionite, organic sulphinates, organic ketoalcoholes or carbohydrates such as glucose, hydroxyacetone, occasionally sulphides and polysulphides are still in use for sulphur dyes [111-115].

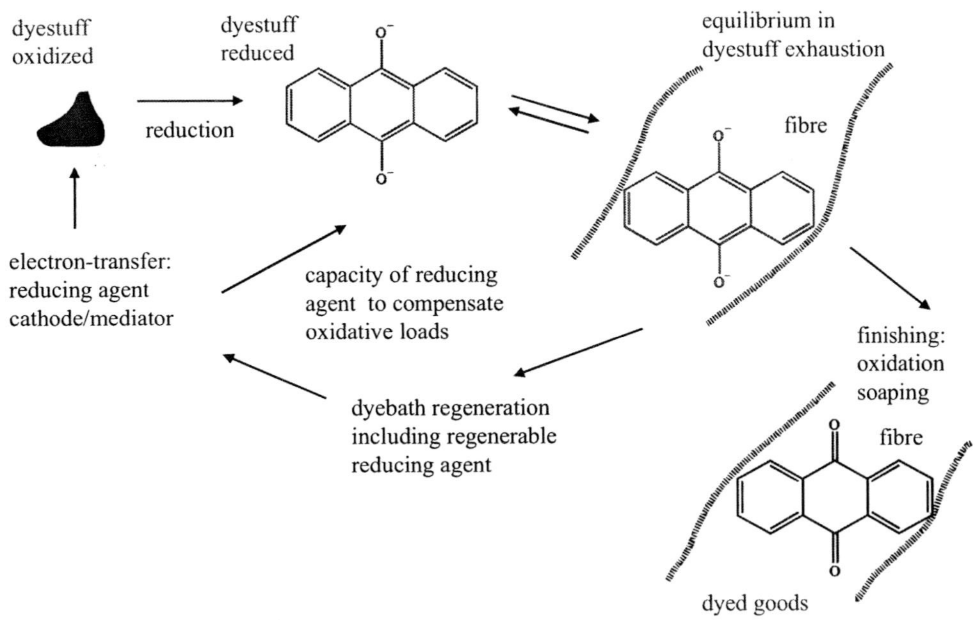

Figure 12. Basic steps in vat dyeing

In Figure 12 an overview about the main steps during the dyeing process is shown. The reducing agents are added to the dyebath in order to reduce the dispersed oxidized dyestuff particles. The reducing agent in vat dyeing represents an electron source for the reduction of dispersed dyestuff particles. In their reduced form the dyestuff molecules get dissolved in the alkaline aqueous dyebath and these forms of the dyestuffs show distinct affinity to cellulosic fibres, which is essential for dyebath exhaustion and consecutive dyestuff fixation.

The final fixation of the dyestuff is achieved by an oxidation step which leads to the insoluble pigment form.

The exhausted dyebath contains the oxidation products of the reducing agents and depending on the dyeing procedure also contains a certain surplus of reducing chemicals, which are required to ensure that the dyes remain in their reduced form during their application.

In particular the surplus of reducing agents and the follow-up products from their use can cause problems in the textile waste-water treatment.

At laboratory scale cathodic reduction of dyestuff pigments could be observed in absence of added mediators however further research is required to distinguish clearly between electrocatalytic hydrogenation and direct reduction of the dyestuff pigment. Indirect electrolysis using so-called mediator systems permits reduction of dispersed dyes at higher current densities and in addition the reduced form of the mediator is able to establish and maintain a certain redox potential in the catholyte. Thus the mediator behaves like a conventional reducing agent which continuously is regenerated cathodically. The requirements for a mediator system thus mainly are defined by the planned application in the dyeing process.

Important requirements for mediators are shown in Table 5.

Table 5. Requirements for mediators, defined by the dyeing process (Potentials vs. Ag/AgCl, 3 M KCl). TEA=Triethanolamine, DGL= D-Gluconate

Dyestuff	Redoxpot	Temp.	pH	Systems	Examples
	mV	°C			
Vat	-1000	80	12 – 14	Iron-amino-complexes	Fe(II/III)TEA
Indigo	-750	40	11 – 12	Iron-carbohydrate-complexes	Fe(II/III)DGL
Sulphur	-500 to -700	95	11 – 13	Anthraquinones	1,2-Dihydroxyanthr., Anthr.-1,5-sulphonate

Important groups of mediator systems for indirect cathodic dyestuff reduction are

- iron-amine complexes e.g. with triethanolamine, HEDTA, bicine as ligands [116-118]
- iron-carbohydrate complexes e.g. with gluconate, hepta-gluconate ligands and selected mixtures [119,120]
- substituted anthraquinones e.g. 9,10-anthraquinone-2-sulphonate, 9,10-anthraquinone-2,5-disulphonate, dihydroxy-9,10-anthraquinones [121]

The main process conditions for a selected dyeing step are defined by the required dyeing conditions and thus research has to start form the recipes used at present. As a second step for a given catholyte an optimisation has to be performed by development of suited electrolysers.

An electrochemical system has to fulfil numerous requirements in order to be used for purposes of textile dyeing:

Requirements defined from the dyeing process:

- sufficient negative reduction potentials(vs. Ag/AgCl, 3M KCl)
    - 450 to -700 mV for sulphur dyes and sulphur vat dyes [122,123]
    - 700 to -800 mV for indigo [124,125]
    - 750 to -1000 mV for vat dyes [126]
- stable reduction conditions in the dyebath
- reproducible process conditions and dyeing results
- similar (or improved) dyeing results compared with the techniques already in use
- low concentrations of chemicals employed and non toxic chemicals
- low costs for chemicals and low losses of chemicals
- economic techniques for recycling of chemicals and washing water, simple waste water treatment
- adaptability to equipment and apparatus in use at present in the dyehouses

Requirements defined from the electrochemical process:

high number of electrochemical cycles of the mediator without cleavage or losses in activity

- high current efficiency and high current density of the reversible redox system
- high reaction rate between reduced mediator and oxidized dyestuff
- sufficient conductivity of the catholyte without changes in dyeing results
- cheap electrode materials for cathodes/anodes
- simply constructed electrochemical cell and easy maintenance
- minimum side reactions e.g. dyestuff destruction due to formation of radicals

CV-experiments permit first estimation about redox potential, current density and reversibility of the investigated redox system.

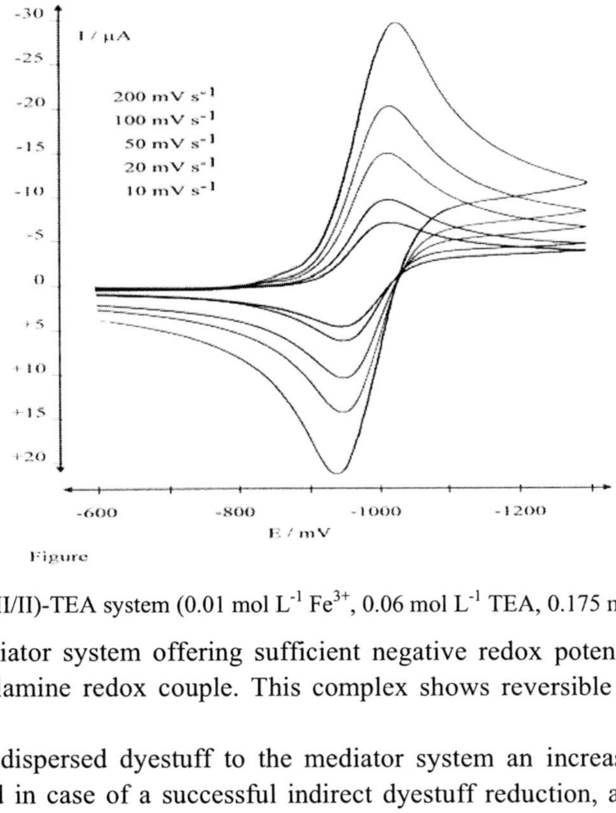

Figure 13. CV of Fe(III/II)-TEA system (0.01 mol L^{-1} Fe^{3+}, 0.06 mol L^{-1} TEA, 0.175 mol L^{-1} NaOH).

A typical mediator system offering sufficient negative redox potential at pH 13 is the Fe(II/III)-triethanolamine redox couple. This complex shows reversible redox behaviour in CV-experiments.

When adding dispersed dyestuff to the mediator system an increase in cathodic peak current is observed in case of a successful indirect dyestuff reduction, at the same time the anodic peak current is reduced because of the redoxidation of the Fe(II)TEA complex. Such behaviour indicates the suitability of a chosen mediator complex. Equations (6) – (8) show the reactions proceeding in indirect cathodic dyestuff reduction when an iron(III)-complex is used, equation (9) – (11) show the mechanism when an anthraquinone (e.g. anthraquinone-1,5-disulphonic acid-sodium salt AODS) is used.

(6) $Fe(III)TEA + e^- \rightleftarrows Fe(II)TEA$

(7) $2\ Fe(II)TEA + Dye \rightleftarrows 2\ Fe(III)TEA + Dye^{2-}$

(8) $Fe(II)TEA \rightleftarrows Fe(III)TEA + e^-$

In selected cases the indirect reduction of the dyestuff can be observed during the CV-scan by photometry in spectro-electrochemical experiments [121].

(9) $AQDS + 2\,e^- \rightleftarrows AQDS^{2-}$

(10) $AQDS^{2-} + Dye \rightleftarrows AQDS + Dye^{2-}$

(11) $AQDS^{2-} \rightleftarrows AQDS + 2e^-$

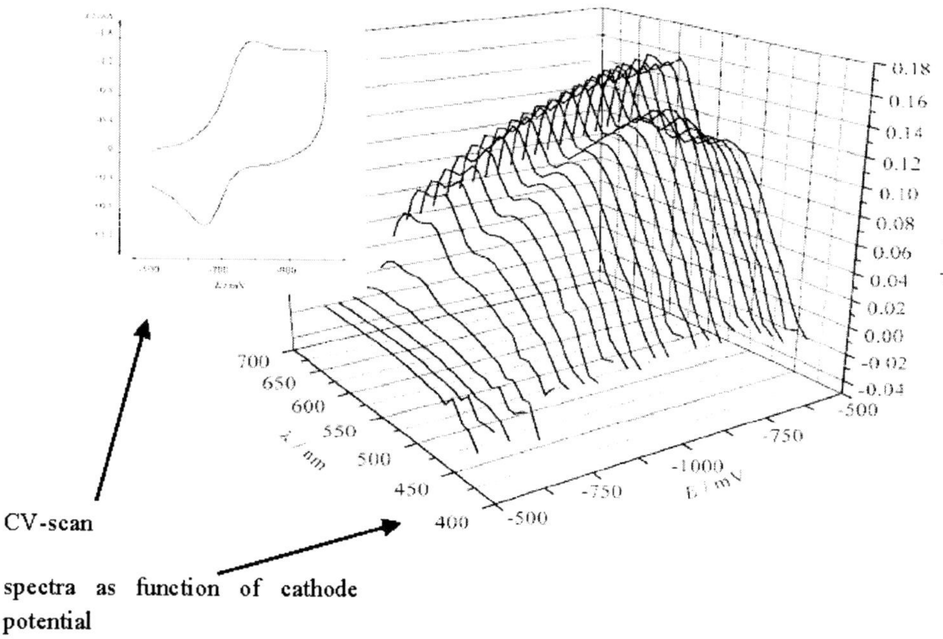

Figure 14 CV and absorption spectra of 0.001 mol L^{-1} AQDS 1.5 in 0,1 mol L^{-1} NaOH in presence of 0.1 g L^{-1} CI Vat Yellow 1.

Both the reduction of AQDS (λ_{max} = 525 nm) and the coupled reduction of the dispersed dye (λ_{max} = 600 nm) can be followed by photometry. According to the CV data the reoxidation of $AQDS^{2-}$ occurs during the reverse scan at approx. -750 mV. At this point the reduction of the dyestuff is stopped completely.

In CV experiments of Fe-TEA complexes in presence of added dispersed dyestuff an increase in the cathodic peak current can be observed which is due to the coupled reduction of dispersed dyestuff. The corresponding increase in current density is not observed to a comparable extent in voltammograms recorded in flow cells when dispersed dyestuff is added to a complex system found suited for dyestuff reduction in CV.

An explanation for this finding can be found when considering the behaviour of the oxidized/dispersed and reduced/dissolved dyestuff in the diffusion layer of the cathode. The oxidized dyestuff is insoluble and enters the diffusion layer relatively slow, while the reduced dyestuff dissolves and leaves the diffusion layer more rapid corresponding to its much higher coefficient of diffusion. At the moment when the dyestuff particles present in the diffusion layer are completely reduced the current density observed in the voltammogram is similar to the current density of the pure mediator system and even the addition of a rather high amount of dispersed dye does not change the current densities to a distinct height. An representation of the concentration profiles of the species present in the diffusion layer in the flow cell is given in Figure 15. The most important species are dyestuff oxidized/insoluble, dyestuff reduced/soluble, Fe(III)-complex and Fe(II)-complex both soluble.

In the CV-experiments the amount of dispersed dye is chosen in such a height that complete dyestuff reduction is not achieved during the CV-scan, so the increase in the cathodic current peak is remarkable and can be used as a measure for the description of the properties of a mediator system. When recording a voltammogram in a flow cell, the concentration of oxidized dyestuff particles present in the diffusion layer is lowered near zero, so the observed voltammograms are close to the results with the mediator system in absence of dyestuff.

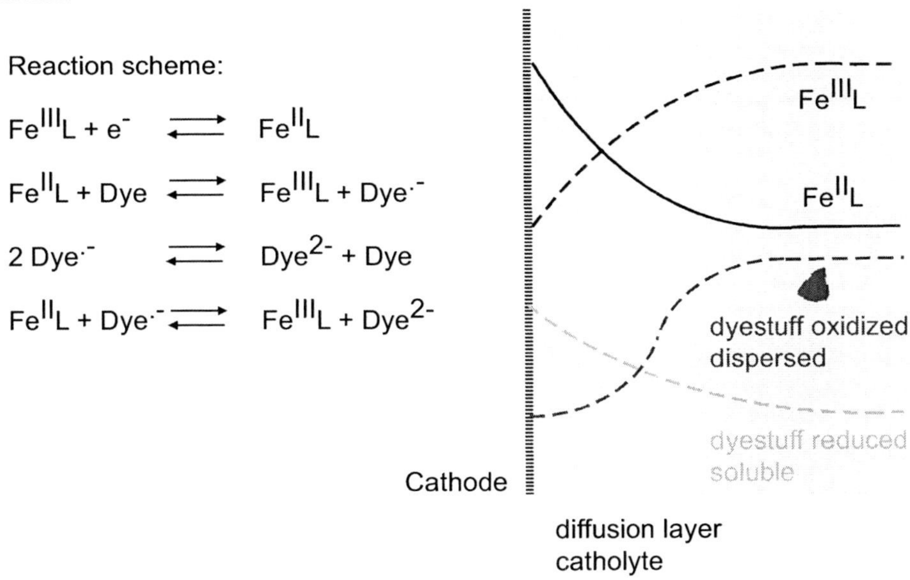

Figure 15. Schematic presentation of concentration profiles in the diffusion layer

While Fe(II/III)-triethanolamine complexes permit to achieve diffusion limited current densities and sufficient negative redox potentials at pH 13-14, Fe(II/III)-carbohydrate complexes e.g. gluconic acid complexes exhibit the greater stability at lowered pH, which is of interest when the dyed material is rinsed and the pH is lowered in the washing bath. As a disadvantage the current density that can be achieved with such carbohydrate complexes is lower compared to the Fe(II/III)-amino complexes.

An optimisation can be achieved when using mixtures of both ligands. Due to the different stabilities of the complexes in their Fe(III)- and Fe(II)-form, the Fe(III)-triethanolamine complex is the main species present in the diffusion layer which is reduced cathodically. The Fe(III)-gluconate complex is present in significant concentrations only when the Fe(III)-concentration exceeds the triethanolamine concentration present in the mixture. In such mediator systems a redox reaction between the Fe(II)-triethanolamine complex formed and the Fe(III)-gluconate present in solution is observed and a remarkable increase in current density is observed. The current density of a given mediator system (fixed iron concentration) thus can be optimised with regard to improved stability and current density. Equations (11) – (15) summarise the reactions occurring during cathodic reduction of mixed mediators.

$$(11)\ Fe(III)TEA + e^- \rightleftarrows Fe(II)TEA$$

$$(12)\ Fe(III)DGL + e^- \rightleftarrows Fe(II)DGL$$

$$(13)\ Fe(II)TEA + DGL \rightleftarrows Fe(II)DGL + TEA$$

$$(14)\ Fe(II)TEA + Fe(III)DGL \rightleftarrows Fe(III)TEA + Fe(II)DGL$$

$$(15)\ Fe(III)DGL + TEA \rightleftarrows Fe(III)TEA + DGL$$

At low scan rates reaction 13 is observed even in presence of TEA at insufficient concentration to complex all iron ions. This indicates a higher stability of Fe(II)DGL. At c(TEA)<c(Fe) a regeneration of the Fe(III)TEA complex according equation 14 is observed. The use of mixtures of ligands permits the successful combination of advantages of two ligand systems (high current density of Fe(TEA), high stability of Fe(DGL) (Figure 16).

A: In presence of high concentrations of TEA (c(TEA) > c(Fe)) the FeIII-TEA complex is present in the diffusion layer and is reduced at diffusion limited current density, Fe(II)TEA formed then reacts with DGL to form Fe(II) DGL

B: when c(TEA) < c(Fe) the current density is limited by the concentration of FeTEA formed. As a considerable part of Fe(III)DGL is present in solution, electron transfer between cathodically formed Fe(II)TEA and Fe(III)DGL proceeds in solution.

Such mixed complexes successfully were used during the scale up of the electrochemical process and led to improved mediator recipes with regard to lower chemical concentration and improved stability during rinsing processes.

An optimisation of the current density is of particular importance because the current densities that can be achieved in pilot scale electrolysers are about 10 - 30 A m^{-2} electrode surface, which is a rather low value.

An increase in the concentration of the mediator system is limited from the point of application in the dyehouse (costs and dyebath composition) so a careful optimisation of the electrolyser has to be performed.

For cathodic indigo reduction the dyeing conditions required different mediator systems to achieve the required redox potential of more negative than -700 mV at a pH-value of 11.5 to 12.0 . These conditions form a fundamental requirement to obtain the desired ring dyeing. At elevated pH-values in the dyebath the penetration of the dyestuff into the yarn increases and the quality of the dyed material is lower.

Iron-amine complexes tend to get unstable at this pH so another group of mediators had to be found. Sugar acids form stable Fe(II/III)complexes at this pH but the redox potential that can be achieved at low ratio c(Fe(II)) : c(Fe(II/III)) is not sufficient negative to permit stable dyebath conditions.

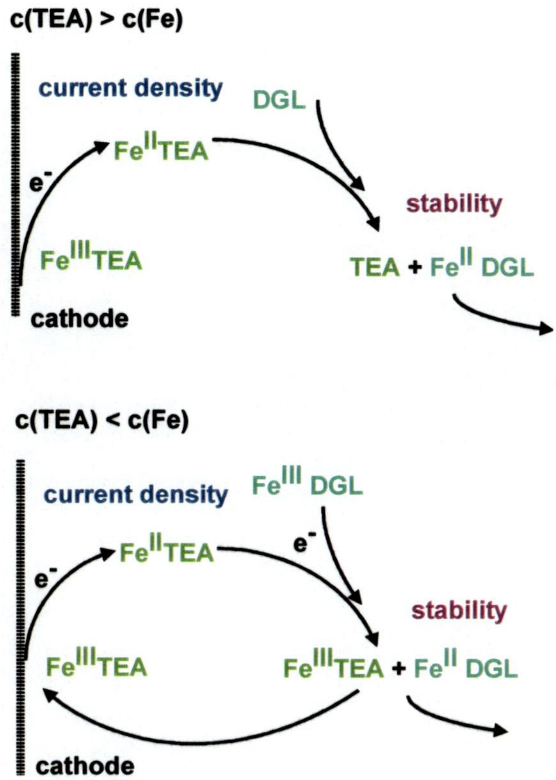

Figure 16. Coupled reactions proceeding in solutions of mixed ligands

For a successful introduction of a complex system the mediator has to reach a reduction potential more negative than -700 mV e.g. -740 mV at lowest possible content of Fe(II)-complex because this amount of complex will be oxidized completely during each air passage of the yarn. This would result in undesired high consumption of Fe(II)-complex, thereby causing an high additional oxidative load and as a result increased cell current would be required to restore the Fe(II)-complex.

An unexpected solution was found with mixed complexes of Fe(II/III)-D-gluconate and Ca^{2+} ions. The uptake of Ca^{2+} into the Fe-D-gluconate complex changes the redox properties of the system. The redox potential measured during constant current electrolysis experiments as a function of the ratio Fe(II)-complex : Fe(II/III)-complex performed with Fe(III)-D-gluconate and Ca^{2+}-Fe(III)-D-gluconate, shown in Figure 17 indicates the formation of redox potentials sufficient negative for indigo reduction even at much lower ratio Fe(II/III) for the case of the CaFeGlc complex compared to the FeGlc complex where a potential more negative than -700 mV is reached at a ratio of 20% Fe(II):Fe(II/III) first. In presence of Mg^{2+} ions the redox potential is not influenced in comparable height.

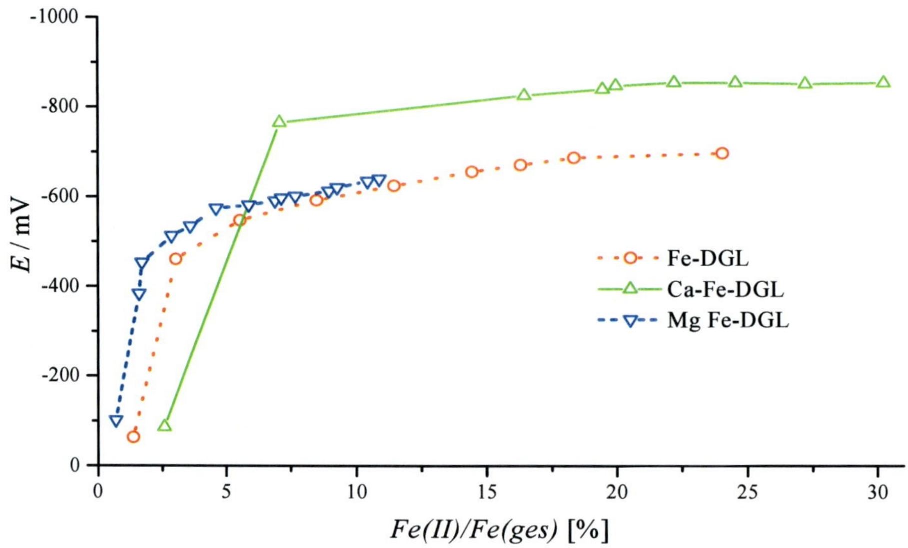

Figure 17. Change in redox potential as a function of the ratio Fe^{II}-complex/$Fe^{II/III}$-complex (total iron) during cathodic reduction of Fe^{III}-complexes. The catholyte composition of experiments A, B and C is given in Table 6.

Table 6. Composition of the catholyts used for the constant current electrolysis tests

Experiment	$c(Fe^{3+})$	c(DGL)	c(NaOH)	$c(M^{2+})$	pH-start	pH-end
	mol dm^{-3}	mol dm^{-3}	mol dm^{-3}	mol dm^{-3}		
A	0.1	0.2	0.50	--	12.49	12.79
B	0.1	0.2	0.59	0.1(Ca^{2+})	12.43	12.73
C	0.1	0.2	0.61	0.1 (Mg)	12.36	12.75

The electrochemical properties of Fe^{3+}-DGL and Ca^{2+}-Fe^{3+}-DGL show distinct differences in their suitability to serve as mediator for indirect cathodic dyestuff reduction. The introduction of Ca^{2+}-ions into Fe^{3+}-DGL complexes leads to formation of a binuclear Ca^{2+}-Fe^{3+}-DGL complex and as a result the formal redox potential shifts approximately –100 mV towards more negative values [127].

Experiments with cyclic voltammetry and batch electrolysis experiments permit to compare the properties of Fe^{3+}-DGL complex and Ca^{2+}-Fe^{3+}-DGL complex to serve as mediator. In presence of dispersed reducible compounds (dyestuff) a catalytic current is observed in the CV-experiments when indirect cathodic dyestuff reduction is achieved. The formation of dissolved reduced dyestuff also can be observed during the reverse scan by an anodic current peak at $(E_p)_a$ of –0.530 V to –0.430 V. This current can be attributed to the reoxidation of reduced dyestuff formed during the cathodic scan.

While in cyclic voltammetry experiments Fe^{3+}-DGL and Ca^{2+}-Fe^{3+}-DGL showed comparable cathodic peak currents $(I_p)_d$, the higher increase of catalytic peak current $(I_p)_c$ can be observed in presence of dispersed dyestuff with Fe^{3+}-DGL. In presence of reducible dispersed dyestuff enhancement factors $(I_p)_c$ / $(I_p)_d$ of up to 1.8 were observed for Fe^{3+}-DGL, while the so-called enhancement factors (*e.f.*) remained below 1.6 using Ca^{2+}-Fe^{3+}-DGL as mediator [128].

Galvanostatic batch electrolysis experiments with a cell current of 0.3 - 0.9 A confirmed the results of the CV-experiments (current density 0.07 - 0.21 mA cm^{-2}, catholyte flow 0.25 cm s^{-1}, inside the porous cathode 0.65 cm s^{-1}). Under the experiment conditions studied, both CI Vat Yellow 1 and CI Vat Blue 5 could be reduced at a cell current of 0.9 A by indirect cathodic reduction with Fe^{3+}-DGL. At the same experimental conditions the Ca^{2+}-Fe^{3+}-DGL mediator failed to establish proper reduction of dispersed dye in the catholyte [128].

4.2. Indirect Electrochemical Reduction of Vat Dyes

For a technical application the total chemical balance of the electrochemical process has to be considered. As can be seen in Figure 18 Na^+-ions are transported through the membrane parallel to the indirect reduction of the vat dye to yield the reduced alkali soluble leuco form. The reduced dyestuff is adsorbed on the fibre and thus transported into the oxidation bath. Dyestuff reoxidation proceeds first in the oxidation bath thus this part of NaOH is not accumulated in the bath. NaOH is formed in the dyebath in case of air oxygen reduction.

In case of $Na_2S_2O_4$ a stoichiometric amount of NaOH effectively is consumed during the reduction step and considerable amounts of sulphite and sulphate are formed in the dyebath and released into the wasted water.

Figure 19 shows a flow scheme of a technical concept for package dyeing of yarn (X-cones). The cell is coupled to the dyeing apparatus and the dyebath is reduced electrochemically. The oxidised dyestuff then is added to the dyebath and gets reduced therein by means of the mediator.

Figure 20 shows the run of an electrochemical dyeing experiment from the point of view of a dyer. The dyeing process normally is described by a temperature-time diagram, in which the important steps are shown. In the case of electrochemical dyestuff reduction two important parameters are shown in addition, the redox potential measured in the dyebath and the cell current applied [129].

In the given example the dyeing apparatus is filled with oxidised dyebath containing the mediator in its oxidised form, redox potential -200 mV. The dyebath circulates through the yarn and through the cell. The cell is working with full power (in this case 45 A). After reduction of the oxygen present in the dyebath has been completed the potential changes towards more negative values.

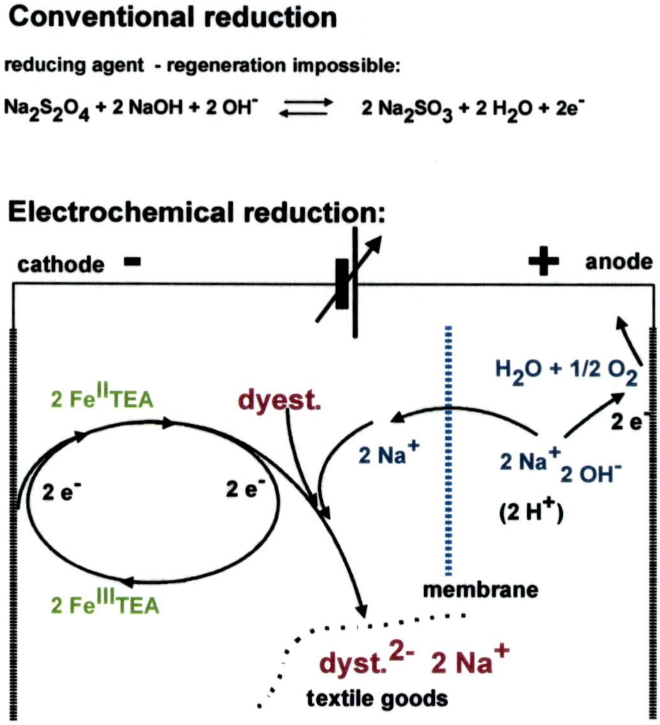

Figure 18. Mass balances in conventional vat dyeing (sodium dithionite and sodium hydroxide) and indirect electrochemical dyestuff reduction

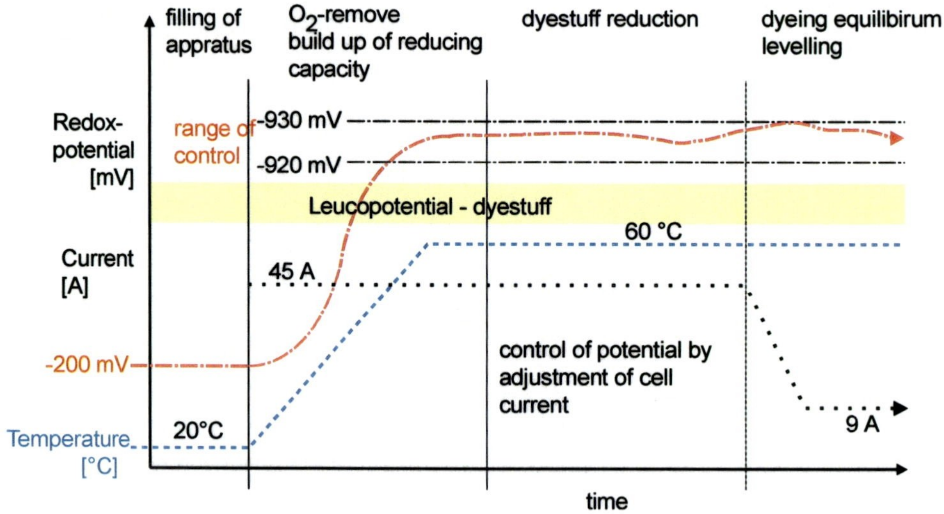

Figure 19. Flow scheme of a electrochemical vat dyeing unit for yarn package dyeing

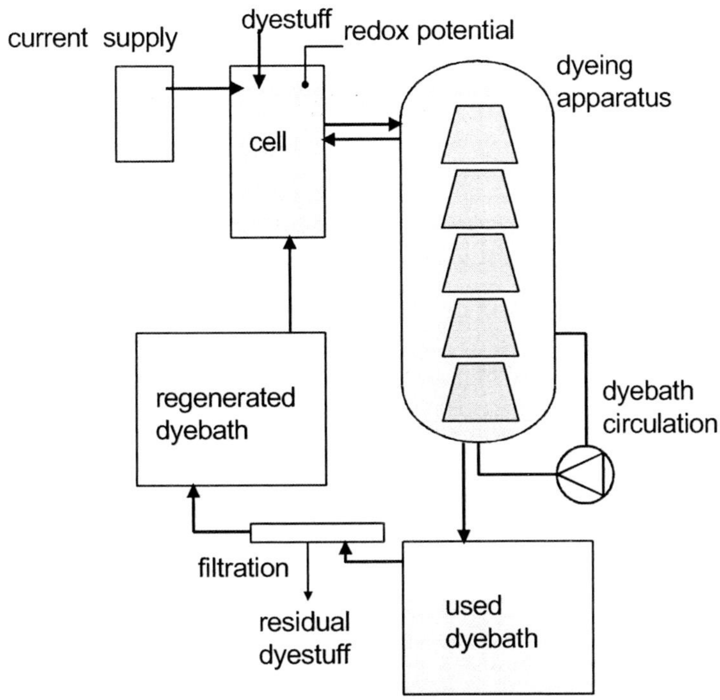

Figure 20. Redox potential in catholyte - dyebath, cell current and temperature during a electrochemical vat dyeing using indirect cathodic dyestuff reduction

The application of mediator systems permits the dyer to adjust the redox potential in the dyebath by control of the cell current. In the given graph the oxidized dyestuff is added to the oxidised dyebath at the begin of the process similar to the pigmentation process in conventional vat dyeing. After turning on the current first oxygen is reduced, then the dyestuff is transformed into the reduced alkali soluble state. The oxidative load introduced by the dyestuff is compensated by the electrochemical reduction equivalents introduced cathodically. At the point of completed dyestuff reduction, the cell current is lowered to the minimum height required to maintain the redox potential within well defined limits e.g. the redox potential is held between -930 mV and -940 mV.

For reproducible dyeings at well controlled conditions in the dyebath the redox potential should be maintained within defined limits, which are determined by type of dyestuff present in the dyebath. Below the so-called leuco-potential of the dye, the risk of partial dyestuff oxidation appears which can cause unequal dyeings. For some blue dyes an upper potential limit exists, above which the dyestuff gets overreduced irreversibly and losses in colour depth and changes in shade are undesired results.

Electrochemical reduction permits improved control of the reduction state of the dyebath. In case of uptake of air oxygen can be compensated by increase in cell current, while in conventional use of reducing agents the risk of uncontrolled dyestuff oxidation can occur.

Figure 21 shows the possibility to compensate an oxidative load e.g. due to air uptake by adjustment of cell current.

In Figure 21 the dependence of redox potential as function of dyeing time in conventional dyeing and EC-dyeing is shown. Oxygen uptake lowers the concentrations of $Na_2S_2O_4$ and

NaOH which then could pass a critical lower limit. In electrochemical dyeing redox potential measurement indicates an additional oxidative load which easily can be compensated by an increased cell current.

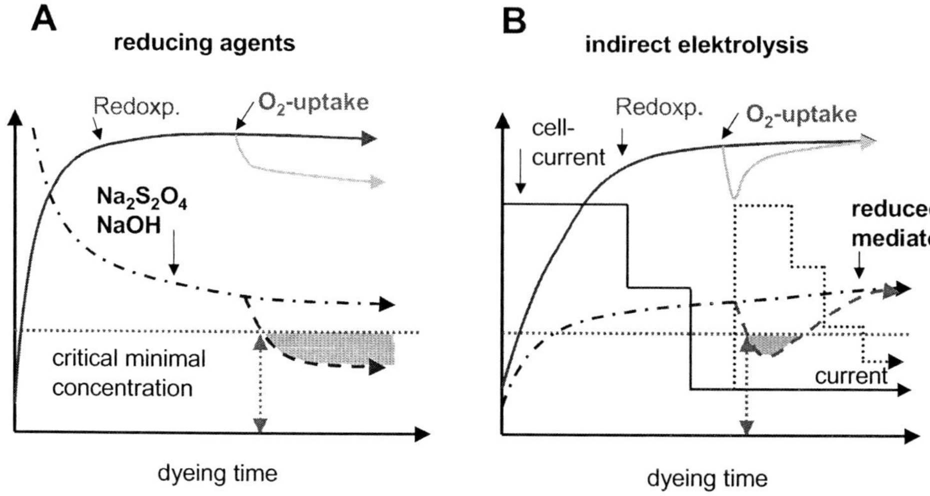

Figure 21. Schematic representation of redox potential, concentrations of $Na_2S_2O_4$ and NaOH respectively reduced mediator and cell current in conventional vat dyeing (A) and in electrochemical vat dyeing (B) [129].

At the end of the dyeing process the dyed material is rinsed with a reduced mediator solution, prepared by cathodic reduction so the dyestuff concentration present in the exhausted dyebath is lowered, which is of importance to meet the fastness properties, particularly the rub-fastness.

After release of this bath into the regeneration tank, the dyeing process is finished as usual.

Besides improved process control and better defined conditions during the dyeing process, savings can be expected from the regeneration of the exhausted dyebath.

After a dyeing process the used dyebath is collected in a storage tank, reoxidised by air oxygen and the precipitated dyestuff is removed by filtration. Soluble components (mediator complex, alkali, auxiliaries) pass the filtration unit and can be stored in the storage tank ready for reuse. The volume of mediator solution lost during handling and processing of the baths has to be replenished.

Table 7 shows savings expected for a 100 kg yarn package dyeing apparatus. The calculation is performed as minimum savings because only process chemicals NaOH and $Na_2S_2O_4$ are considered for the comparison with the EC-reduction. Losses of 15% of the dyebath volume are considered to appear in the case of EC-reduction due to losses during dyebath filtration and losses from the transportation of a residual amount of dyebath with the wet yarn into the washing/rinsing baths.

Table 7. Chemical consumption and savings, IN-process, 100 kg Co, liquor ratio 1:10

Colour depth	EC-reduction 0.3 - 4 %		Dithionite 0.3 %		2 %		4 %	
	conc.	weight	conc.	weight	conc.	weight	conc.	weight
Dyeing:								
complexing agent	13.6 g L^{-1}	2050 g						
dithionite			3 g L^{-1}	3000 g	5 g L^{-1}	5000 g	8 g L^{-1}	8000 g
NaOH	32 ml L^{-1}	2100 g	14 ml L^{-1}	6160 g	19 ml L^{-1}	8360 g	24 ml L^{-1}	10560 g
Reduct. rinse:								
dithionite	-	-	-	-	2 g L^{-1}	2000 g	2 g L^{-1}	2000 g
NaOH	-	-	-	-	2 ml L^{-1}	880 g	2 ml L^{-1}	880 g
Total amount		4150 g		9160 g		16240 g		21440 g
Savings				55 %		75 %		80 %
Water - Dyeing		150 L		1000 L		2000 L		2000 L

only process chemicals considered,
losses during recycling of mediator c. 15 % of dyebath volume

In a technical research cooperation a full scale installation of an electrochemical cell coupled to a dyeing apparatus was planned, built up, and brought to successful operation. The dyeing apparatus had a capacity of 110 kg cotton yarn per batch.

The installation included:

- 500 A multi-cathode electrolyser for indirect cathodic dyestuff reduction
- ultrafiltration unit for recovery of the dyebath containing the mediator solution
- installation for storage of used dyebath and regenerated mediator solution
- control devices including steering programs.

Characteristic parameters of the installation are given in the Table 8.

A key process for the expected ecological profile and the economic benefits is the regeneration of the used dyebath by nanofiltration. For this purposes a nanofiltration unit (filtration capacity 1 m^3 h^{-1}) for dyestuff removal had to be designed and installed. Storage tanks to collect used dyebaths before filtration and for storage of recycled liquors were required (each 3 m^3 capacity).

4.3. Indirect Reduction of Sulphur Dyes

As already described in part 1.1. of this chapter sulphur dyes can be reduced cathodically in absence of added mediator systems. Due to the particular redox behaviour sulphur dyes can pass several stages until full reduction is achieved. The dyestuff already dissolves in partially reduced state, thus the dyestuff is able to serve as mediator itself. This technique is advantageous for processes performed at rather high dyestuff concentration e.g. 10 – 100 g L^{-1} Sulphur Black 1.

To achieve a certain reduction capacity in a dyebath and to maintain proper state of reduction mediator systems can be used in addition, particularly when lower dyestuff concentrations are used.

Table 8. Relevant data for the full scale electrochemical vat dyeing unit

Electrolyser	Material
2x2x10 - cathode electrolyser 2 symmetric electrolyser units 2 cathode groups per unit 10 cathodes per group	stainless steel mesh
catholyte	Fe(III/II)-complex, NaOH
temperature	20 – 80°C
pH	12 – 13
redox potential	-920 to -940 mV (vs. Ag/AgCl, 3 M KCl)
dyestuff concentration	<10 g L^{-1}
separator	cation exchange membrane
anode	steel expanded mesh
anolyte	1 – 2.5 M NaOH
cell current	500 A
cell voltage	<5 V
catholyte volume	1200 L
production capacity (cotton yarn)	112 kg
nanofiltration unit capacity	1 m^3 h^{-1}

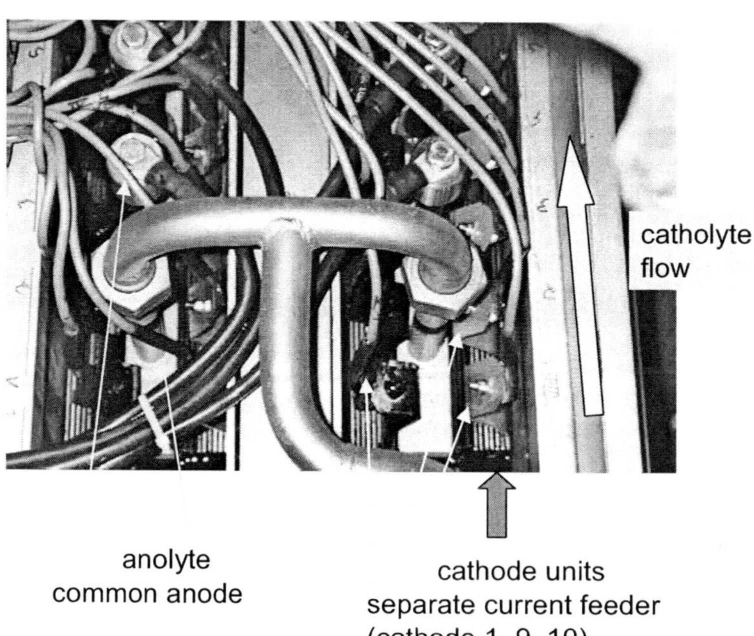

Figure 22. Top view of electrolyser (the arrow indicates the flow of the dyebath through the cell)

Sulphur dyes require reduction potentials between –500 to –700 mV (vs. Ag/AgCl, 3 M KCl reference) which is a lower value compared to vats, thus redox couples based on Fe(III/II)-complexes could serve as mediators.

Detailed experimental studies showed that the sulphide content present in most sulphur dyes can cause precipitation of FeS from the iron-complexes, particularly on the cathode surface, where highest concentrations of Fe(II) are present. The high dyebath temperature of up to 95 °C also favours use of organic anthraquinone (AQ) based mediators.

Important examples are 9,10-anthraquinone-2-sulphonate, 9,10-anthraquinone-1,5-disulphonate, 1,2-dihydroxy-9,10-anthraquinone. Even at concentrations of less than 1 g L^{-1} anthraquinoid mediators successfully can be applied in electrochemical dyeing processes. Depending on the formal redox potential of the anthraquinone used redox potentials between –600 mV (9,10-anthraquinone-2,5-disulphonate) and –780 mV (1,2-dihydroxy-9,10-anthraquinone) is achieved in the dyebath. However a considerable difficulty to be considered for general application of AQ-based mediators is their colour which could disturb the final shade.

4.4. Indirect Reduction in Indigo Dyeing

Despite the fact that indigo (Vat Blue 1) can be considered to be a vat dye the application of this dye is focussed on production of denim. This process is performed on specialised equipment and at dyeing conditions designed to obtain minimum penetration of dye into the yarn (ring-dyeing). Main differences to the conventional vat dyeing procedures can be summarised as following:

- lower pH of the dyebath which is around 11.5 to 12.0 (limited concentration of ground electrolyte)
- dyeing at low temperature (25-30°C)
- continuous process based on a dyebath with continuous replenishment.

Data describing relevant parameters and indicating the enormous dimensions of a typical dyeing unit are shown in Table 9.

The dyeing at this low pH-value required the development of special mediator systems. As mentioned above binuclear complexes of the general type Ca^{2+}-Fe^{3+}-D-gluconate exhibit more negative formal redox potential than the mononuclear Fe^{3+}-D-gluconate complex.

Based on laboratory scale experiments and scale up tests a full scale dyeing test could be done using a 1000 A multi cathode electrolyser specially designed for the process [130].

A flow scheme of an electrolyser for continuous cathodic reduction of the indigo dyebath including a membrane unit (nano- and ultrafiltration) for recovery of the diluted dyebath is shown in Figure 23.

Instead of continuous addition of hydrosulphite and caustic soda, which causes long time variations in the composition of the dyebath, cathodic reduction is applied to compensate the oxidative load brought in by the air oxidation step.

Table 9. Working conditions and production data for a full scale indigo dyeing range

Production rate:	15,000 kg cotton/day
	11.9 kg min^{-1}
hours of operation	21 h day^{-1}
warp speed	35 m min^{-1}
depth of shade	2% indigo
Consumption of chemicals:	
reducing agent $Na_2S_2O_4$	50 - 126 kg day^{-1}
	40 - 100 g min^{-1}
water consumption	3 - 5 L kg^{-1}
	45 - 74 m^3 day^{-1}
Composition of dyebath = catholyte	
wetting agent	0.5 g L^{-1}
pre-reduced indigo	1 - 4 g L^{-1}
NaOH to maintain pH	11.5 - 12.0
temperature	20 - 30°C
redox potential	<-700 mV
Cell dimension (Faraday, current efficiency 80 %):	
minimum	940 A
expected: up to	2300 A – 5000A

In each case of development particularly when a nearly closed loop in waste water regeneration is attempted, a careful consideration of the mass balances is necessary to avoid the formation of components that will accumulate in the system.

The mass balances of an electrolyser for indigo dyeing are similar as already presented for vat dyeing (Figure 18).

The reduction of indigo and the following oxidation by air oxidation produces 2 OH^- ions which are released into the dyebath during the next dip.

The use of $Na_2S_2O_4$ consumes NaOH and forms Na_2SO_4 and Na_2SO_3. Using $Na_2S_2O_4$ 4 OH^- are consumed during its use as reducing agent, when air oxygen is reduced 2 OH^- are consumed. Both the Na_2SO_4 and Na_2SO_3 cause problems to meet the waste water limits existing in many countries [124].

In addition the Na_2SO_4 accumulates in the dyebath thereby changing the affinity of the indigo to the cotton yarn. As a result the problem of insufficient long time stability of the dyeing process is a well known problem. The so-called "ending" is the result of the continuous (but not constant) addition of reducing chemicals (salts) to the dyebath.

An important advantage of the indirect electrochemical reduction in denim production is the expected higher long-time stability of the composition of the dyebath, particularly when an acidic anolyte is used, because the overall product resulting from the reduction of oxygen then is water.

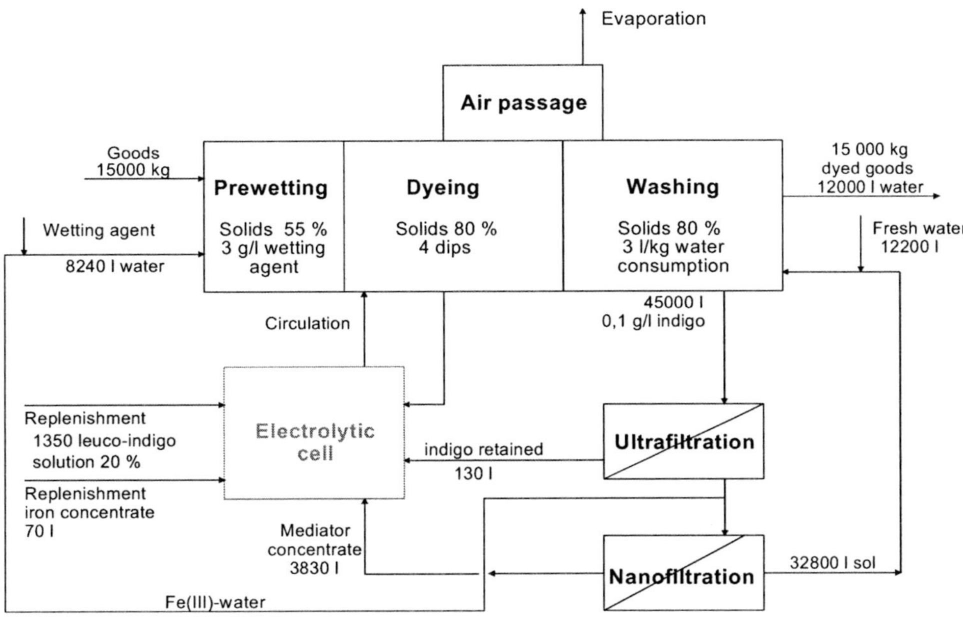

Figure 23. Scheme of a (planned) full scale indigo dyeing range using indirect cathodic dyestuff reduction including dyebath recovery by membrane filtration

5. Cells and Equipment

5.1. Cell Design

In contrast to homogeneous reactions in solutions and gas phase electrochemical reactions already take place under very anisotropic conditions. The simplest electrodes are plate or cylinder electrodes. Due to their low production costs they have been frequently used in industrial application. Because the rate of conversion in electrochemical cells is proportional to the electrode surface and the concentration of the electrochemical active species as well in the direct as indirect reduction of vat dyes is already low or moderate the utilisation of cells owing three-dimensional electrodes or electrodes with a high surface is necessary [131,132].

The space-time yield, however, determines the specific investment and the process costs and is proportional to the product of the specific electrode area and the current density (eqn. 16)

$$STY = a_e * \frac{j * M_P * \beta}{z * F} \tag{16}$$

a_e specific electrode area

β current efficiency

j current density

M_P molar Mass of the product

Typical cells having a three-dimensional electrode are electrochemical cells with porous electrodes, packed-bed electrodes or fluidised-bed electrodes. Unfortunately in these three-dimensional cells not only the mass-transfer problem but also the current-distribution problem must take into account. In fact the current and potential distributions mainly depend on the cell geometry, especially on the relative arrangement of the three-dimensional working-electrode and the counter-electrode.

In general there are three basic principles to achieve an optimum of space-time yield in electrochemistry:

1) mass transfer and the current density can be enhanced by setting the electrodes in motion or by using turbulence promoters (e.g. Chemelac cell).
2) Large electrode surfaces can also be accommodated in a small reactor volume, resulting in developments such as the multicathode cell of DeNora and an electrolytical cell containing multiple expanded electrodes.
3) By using three-dimensional electrodes such as porous flow-through cell, the packed-bed cell, the fluidised-bed cell and the rolling tube cell, high mass transfer coefficients and large specific electrode areas are realised.

In the three-dimensional electrode the penetration depth of the current is restricted in the direction parallel to current flow. The penetration depth of the current density increases as the concentration, which is proportional to the limiting current density, decreases.

Despite of the fact that Rys et al. have used fluidised- and packed-bed electrodes for the reduction of vat dyes and indigo there are some disadvantages of these cells.

The frequently insufficient hydrodynamic approach flow and the unfavourable distribution of potential are the reasons that the efficient electrode area is only partly useable for electrochemical conversions. Often a decline of current density can be observed in the surface-layer that have a certain distance to the counter-electrode.

To avoid these problems Bechtold et al. have applied the multi-cathode cell to reduce sulphur, vat and indigoid dyes [133,134].

While the cell construction used in the in-situ formation of oxidants (e.g. hypochlorite) follows the usual concepts of a undivided parallel plate cell, the general design of the electrolysers for the application in textile processing is determined by comparable low current densities that can be obtained in the systems.

Due to the limited concentration of the electroactive species / redox couple present in the electrolyte typical current densities measured in flow cells are around 20 A m^{-2} for 0.01 M Fe(III/II)-triethanolamine and 8 A m^{-2} for 0.001 mol L^{-1} AQDS1,5. At increased mediator concentrations e.g. 0.024 M Fe-complex current density can be increased to 40 - 50 A m^{-2}.

Cell optimisation can be done with regard to electrode material (steel, glassy carbon, carbon felt), electrolyte flow and general cell construction [133,135].

As the concentration of electroactive species present in solution either is limited by the process costs (e.g. in dyeing procedures, bleach processes), the losses in the effluents or by the electrolyte itself (dye concentration in dyestuff decolourisation), optimisation has to be performed mainly in the direction of improved cell design.

For these applications divided cells are used and as current density is limited only on one electrode (e.g. cathode) development of cell concepts with high cathode surface was necessary. To limit cell costs membrane area and anode area have to be kept low. Based on porous three dimensional electrodes two general concepts are in use at present:

Multi cathode electrolysers and multi-layer-expanded metal cathodes.

The concepts have been extensively tested both in laboratory scale and in full scale operation.

In multi cathode electrolysers current is fed into a set of parallel isolated cathodes by separated power supplies which are connected to a common anode. Electrolyte flow through the cathode stack is adjusted in such height that concentration polarisation is low. [134,136,137].

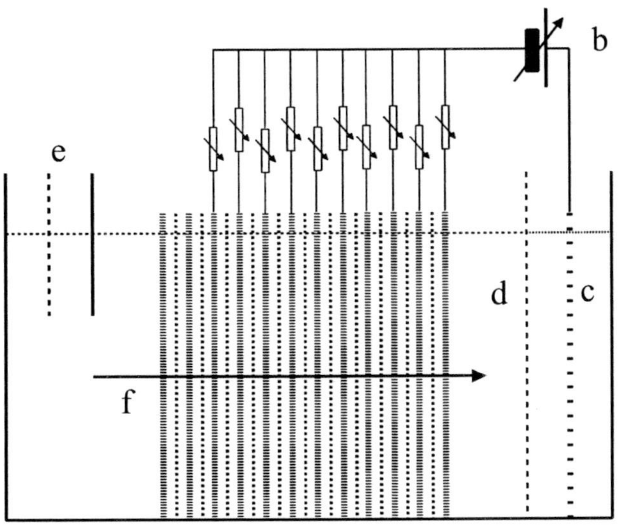

Figure 24. Scheme of the multi-cathode electrolyser (a) cathodes, (b) power supply, (c) anode, (d) cation exchange membrane, (e) measurement of redox potential and temperature, (f) catholyte circulation.

In a laboratory multi cathode cell using 10 stainless steel fabric cathodes (each 0.04 m^2 geometric area, surface area 0.096 m^2) were combined to a cathode stack with a total area of 0.4 m^2 (surface area 0.96 m^2). Using a 0.012 M iron-complex solution permits a current density of 20 A m^{-2} and a cell current 8 A, compared to 0.8 A when a single cathode unit is used.

A full scale multi cathode electrolyser for 500 A cell current is in use for electrochemical reduction of vat dyes. This flow through cell is coupled to a 112 kg dyeing apparatus and is used for test dyeing in full scale. Figure 25 shows a picture of the electrolyser.

The divided cell is built up by two identical elements, each consisting of centre anode and 10 isolated cathodes on each side of the anode.

The cell voltage as function of the position of the cathode unit measured at 400 A total cell current is shown in Figure 26.

Figure 25. 500 A multi cathode electrolyser for indirect electrochemical reduction of vat dyes (arrow indicates flow direction of catholyte/dyebath inside the cell).

Multi cathode electrolysers permit high specific cathode area however proper design is essential to obtain the desired current efficiency. Attention has to be given on the potential drop in the electrolyte particularly within the cathode stack. At high cell current the cathode units at the front of the cathode stack can enter into bipolar behaviour. The current flow in the catholyte then causes a potential difference between front and back side of a cathode unit in such height that the electrodes work as bipolar electrodes. In the most simple case the efficiency of the cell is decreased. In the case of vat dye or indigo reduction reoxidation of already reduced dyestuff on the back-side of a cathode blocks the catholyte flow through the porous electrode.

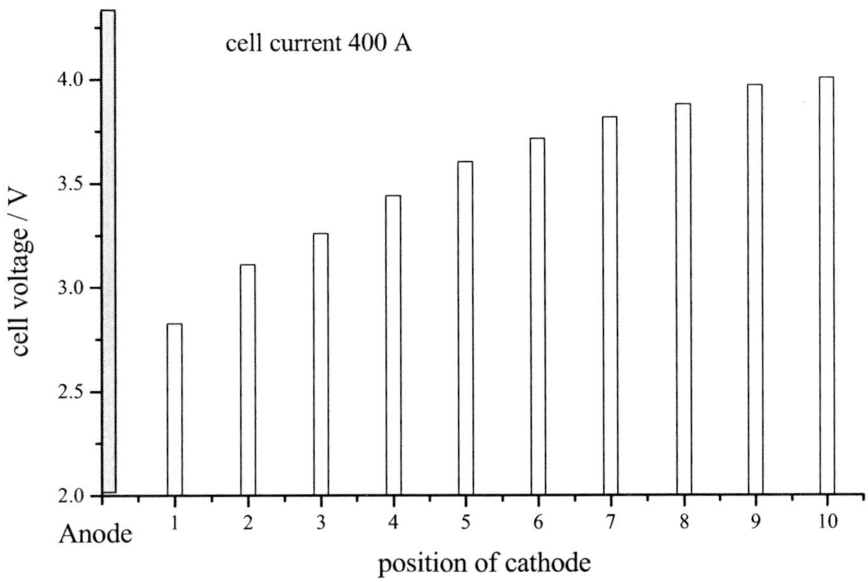

Figure 26. Cell voltage as function of the position of the cathode units; position (1) shortest distance to anode, position (10) cathode unit at back side of cathode stack. Cell current of 400 A. Experimental conditions: Anolyte 1 M NaOH, catholyte 33 ml L^{-1} mediator concentrate (DyStar), 15ml L^{-1} NaOH 50°Bé, 0.5 ml L^{-1} wetting agent 3.75 g L^{-1} vat dye.

In the case of indirect reduction of dispersed vat dyes a determination of the acceptable potential drop ΔU is possible with use of voltammetric experiments [133,138]. The potential interval in which the current plateau attributed to the desired electrode reaction is observed indicates the maximum potential drop which might occur along a cathode.

An estimation of the potential drop in a cathode can be made according equation (17) in Figure 27.

An optimisation of the construction of the cell is the preferable strategy, because in the major part of textile processes modification of the catholyte (dyebath) is hindered by requirements defined by the dyeing process.

An optimised multi-cathode stack is also shown in Figure 27. As a result of the increased current flow in the electrolyte through the electrodes near the anode the thickness of the cathodes has to be decreased.

As mentioned above the decline of the current density depends on the distance of the electrolytical cathode-layer or particle to the counter-electrode. The multi-cathode cell can bypass this problematic aspect by the combination of several plan-parallel and liquid permeable cathodes that are isolated to each other.

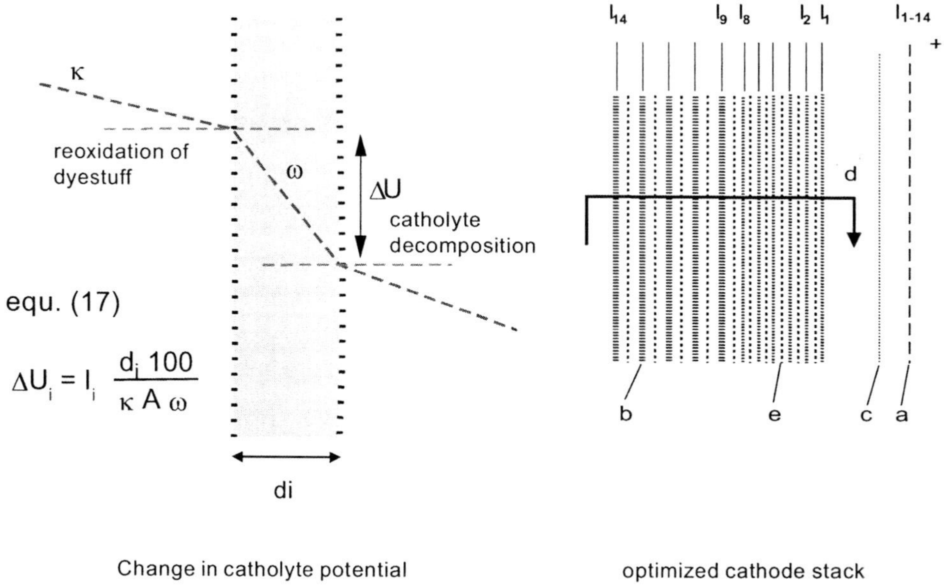

Figure 27. Equation for calculation of potential drop in the electrolyte within a cathode unit of a multi cathode electrolyser and optimised cathode stack (I_1 – I_{14} current fed into cathode unit i, a anode, b cathode unit, c cation exchange membrane, d catholyte flow, e insulator).

ΔUi potential change in the cathode i (V)
Ii current flow in the electrolyte (A)
di thickness of electrode (cm)
κ conductivity of electrolyte (mS cm-1)
A front area of cathode stack (cm²)
ω part of cross section of the electrolyte that can be used for current transport within the cathode (porosity, %)

Due to the fact that the more distant cathode of the anode show an increasing ohmic resistance this effect has to be compensated by feeding a higher voltage for these electrodes. In general this can be reached by a separated current supply for each cathode. Thus the instrumental expenditure is very high and the separated current supply requires either several power suppliers or by using one power supplier the in-line arrangement of several different adjustable resistances. In the case of the use of external adjustable resistances the consumption of electrical energy is more significant because a part of the used direct current is transferred in heat. Furthermore the construction of the electrolytic cell is very sophisticated. The separated cathodes have to be electrically isolated arranged and supplied with different power connections. Due to the vertical electrolyte-flow the mass-transfer is not optimised. Thus a high circulating amount of electrolyte is necessary to achieve sufficient electrolyte flow along the transversely flowed electrode-surfaces. Because of the concept and properties of this cell it is not possible to construct bipolar electrolysers to realise a most efficient technical solution for industrial approaches.

A solution for this issue was found by EUT and DyStar [139]. They have recently developed a new divided electrolytic cell containing multi-layer-expanded metal cathodes. This cell consists on multiple expanded metal cathodes, which are separated via spacers and are arranged in the catholyte space. Using this approach monopolar and bipolar electroysers are possible.

The cell distinguishes itself by the fact that the cathodes are designed as multi-expanded metal electrodes. The electrodes consist of expanded metal-layers that connected *via* internal resistors are flowed through by the electrolyte. In general 4 up to 12 expanded metal layers are combined to one multi-expanded metal electrode which is in contact to a charged plate electrode.

In this connection interlayers consisting of porous electrode material (e.g. metal foams) can be used.

Based on the longitudinal flow of the electrolyte optimal conditions for a good mass transfer can be achieved. Due to the continuous change of direction of flow and the alternate narrowing and expansion of the flowed cross section micro-turbulences occur facilitating a very advantageous mass transfer. Because of the preferred strength of the expanded metals of 1.5 to 3 mm there are relatively great distances between the charged plate electrode and the most distant expanded metal layer (about 30 mm using 10 to 12 layers). In dependence of electrolytic conductivity of the electrolyte this causes potential differences between 0.5 to 2V and a decrease of the current density with respect to the distances of the several layers to the counter electrode.

Using the internal resistors between the individual layers this drop of current density can be bypassed so that a corresponding difference of potential is formed between the plate electrode connected with the power supply and the adjacent expanded metal layers. This results to a significant simplification of the power supply and the measurement and control devices.

On the contrary to the multi-cathode cell the possibility to construct very efficient bipolar electrolysers is a significant feature of this cell concept.

The application of this cell has proven very good performance both in direct and indirect cathodic reduction of vat, sulphur and indigoid dyes. At optimised conditions excellent current efficiencies near the theoretical values were observed. At the moment first experiences are gained on large scale using this concept and this electrolyser cell is on the threshold of the industrialisation.

5.2. Process Control

Based on basic investigation of a selected system by cyclic voltammetry the applicable current density is determined by voltammetry in flow cells and by experiments with determination of current efficiency. Particularly in the case of three dimensional electrodes, multi-layered electrodes and multi-cathode electrolysers larger scale cells are driven at galvanostatic conditions.

The maximum height of applied cell current is limited by type and concentration of the electrochemical system used and the construction of the electrolyser. Depending on the textile process a suited process control is required:

- for indirect electrolysis in vat dyeing and indigo dyeing a certain redox potential in the electrolyte has to be adjusted. As indicated by Figures 20 and 21 redox potential measurement in the catholyte (dyebath) is used as parameter to define the cell current. As long as the selected potential is has been reached in the dyebath, the cell operates at planned maximum current (e.g. 500 A). When the selected range of dyebath potential has been reached, the cell current is used for regulation of the redox potential. From practical experiences the basic consumption of reducing equivalents required to maintain a certain redox potential should not exceed 10% (e.g. 50 A) of the maximum current (500 A) of a cell.

- an installation for cathodic decolourisation of waste water should operate fully automated. The composition of the waste water with regard to the present dyes varies in concentration and type of dyestuff. In presence of resistant dyes e.g. phthalocyanine based dyes or anthraquinoid reactive dyes, photometric sensors would hinder release of treated baths because a set limit would not be achieved. Thus the technical process is running as a batch process with fixed charge flow per m^3 of coloured waste water. As a result the cell operation is extremely simple and risk of failure is minimal.
- for in-situ chlorine bleach operation both control techniques can be applied. Either measurement of redox potential and steering of bleach process by regulation of cell current is possible or a certain amount of oxidant is formed before the bleach step or during the bleach by a preset charge flow through the cell.

6. Ecological Considerations

6.1. Process Requirements

Any process introduced into textile dyehouses at present will have to fulfil the existing regulations waste water from textile processing [28,32]. Selected parameters are given in Table 11. Besides general parameters like pH, temperature, COD (chemical oxygen demand), BOD (biological oxygen demand), attention has to be paid on heavy metals released with the effluents. A number of Co-, Mn-based complexes could be considered to serve as mediators for indirect electrolysis however the existing limitations for the concentration of such metals in the wasted water will make introduction of such systems into full scale operation very difficult.

6.1. Requirements for Textiles

A considerable part of the textiles are used in direct contact with human skin. To avoid irritation, sensitisation or even potential risk of cancer any substance which is applied in textile processing has to be considered with great care.

During the last decade numerous limits have been defined for hazardous substances on textiles, including selected banned azo-dyes, heavy metal content and formaldehyde concentration on textiles. As a result any chosen mediator system planned to be used in indirect electrochemical processes for dyeing, bleach etc. has to fulfil these requirements and limits which are widely accepted at present. This is of particular importance when organic mediator systems or metal complex-based mediators which contain heavy metals other than iron are chosen and discussed as technically feasible systems.

Table 10. Representative limits defined for release of textile waste water

Limits for emission		release into river	release into CWWT
General parameters			
temperature	°C	30	40
toxicity		< 2	no hindrance of biodegradation
pH		6.5-8.5	6.5-9.5
colour, spectral coefficient of extinction:			
436 nm (yellow)	m^{-1}	7.0	28.0
525 nm (red)	m^{-1}	5.0	24.0
620 nm (blue)	m^{-1}	3.0	20.0
Inorganic parameters			
aluminium	mg L^{-1}	3	limited by filter residue
chromium total	mg L^{-1}	0.5	1
iron	mg L^{-1}	2	limited by filter residue
cobalt	mg L^{-1}	0.5	0.5
copper	mg L^{-1}	0.5	0.5
tin	mg L^{-1}	1	1
sulphate (as SO_4)	mg L^{-1}	-	200
Organic parameters			
TOC (total organic carbon as C)	mg L^{-1}	50	>70 %biodegradation
COD (chemical oxygen demand as O_2)	mg L^{-1}	150	>70% biodegradation
BOD_5 (biological oxygen demand) O_2)	mg L^{-1}	20	-

REFERENCES

[1] Grotthus, C.J. *Ann. Chim. Phys.* 1807, *63*, 18.

[2] Goppelsroeder, F. *Studien über die Anwendung der Elektrolyse zur Darstellung, zur Veränderung und zur Zerstörung der Farbstoffe, ohne oder in Gegenwart von vegetabilischen oder animalischen Fasern,* Reichenberg 1891.

[3] Goppelsroeder, F. *Über die Darstellung der Farbstoffe, sowie deren gleichzeitige Bildung und Fixation auf der Faser mit Hilfe der Elektrolyse* Reichenberg 1885

[4] Goppelsroeder, F. *Dinglers Polytech J.* 1882, *245*, 225.

[5] Goppelsroeder, F. *Dinglers Polytech J.* 1884, *251*, 465.

[6] Goppelsroeder, F. *Dinglers Polytech J.* 1884, *253,* 245.

[7] Goppelsroeder, F. *Dinglers Polytech J.* 1884, *253*, 381.

[8] Goppelsroeder, F. *Chem. Ztg.* 1893, *17*, 1633.

[9] Wartha, V. *Chem. Ztg.* 1884, *8*, 431.

[10] Mullerus, J. *Chem. Ztg.* 1893, *17*, 1454.

[11] Christensen, M.; Horstmann, G. *Melliand Textilberichte* 1987, *68*, 50-51.

[12] Phillips, D. *J.S.D.C.* 1996, *112*, 183.
[13] Shankarling, G.S.; Paul, R.; Jayesh, M. *Colourage*, 1996, *43/8*, 47-50, 52-54.
[14] Shankarling, G.S.; Paul, R.; Jayesh, M. *Colourage,* 1996, *43/9*, 57-58, 60-62.
[15] Bechtold, T.; Burtscher, E.; Bobleter, O. in *Recent Research developments in Electrochemistry;* Pandalai, S.G. (Ed.); 1998, *1*, 245-264, Transworld Research Network, Trivandrum – 695008, India.
[16] Bechtold, T.; Berktold, F.; Turcanu, A. *J.S.D.C.* 2000, *116,* 215-221.
[17] Brunner, H.; Richter, R.; Gauweiler, B.; Bechtold, T. *manuscript in preparation.*
[18] Bechtold, T.; Burtscher, E.; Turcanu, A. *Text. Chem. Color.* 1998, *30/8*, 72-77.
[19] Bechtold, T.; Burtscher, E.; Turcanu, A. *J. Appl. Electrochem.* 1998, *28*, 1243-1250.
[20] Bechtold, T.; Burtscher, E. 1998 *Pat. Appl.* PCT/AT 98/00207.
[21] Brunner, H.; Bechtold, T.; Richter, R.; Schlerka, W.; Grütze, J. 2004 Pat. Appl. DE 102004040601.4
[22] Lange, O. *Die Schwefelfarbstoffe, ihre Herstellung und Verwendung,* 1925, 2nd Ed. , Spamer, Leipzig.
[23] Zerweck, W. ; Ritter, H. ; Schubert, M. *Angew. Chem.* 1948, *60*, 141-147.
[24] Frind, H.; Held, C.; Aman, H. 1969, *Offenl. DE 1906083.*
[25] Bechtold, T.; Burtscher, E.; Turcanu, A.; Berktold, F. *Melliand Textilber.* 2000, *81*, 195-200.
[26] Bechtold, T.; Turcanu, A.; Burtscher, E.; Bobleter, O. *Textilveredlung* 1997, *32*, 204-209.
[27] Bechtold, T.; Schrott, W.; Hüls, T.; Muche, M.; Wendt, B. 2002 *Pat. Appl. PCT/EP 03/08050.*
[28] Regulation of the Austrian Ministry of Agriculture – Limitation of the emission released from textile finishing and textile treatment plants, Austrian BGBL, 207 Vienna, Austria (1992).
[29] T. Bechtold T.; Mader C.; Mader J. *J. Appl. Electrochem.* 2002, *32*, 943-950.
[30] Bechtold T; Mader C.; Mader J. *Melliand Textilberichte,* 2002, *83*, 361-364.
[31] Levafix PN, Procion P, Procion PX livens up your reactive prints, Application brochure, 05.1032-00, DyStar, Frankfurt.
[32] H. Schönberger H.; Kaps U. *Reduktion der Abwasserbelastung in der Textilindustrie*; Umweltbundesamt: Berlin, 1994.
[33] Kolb M.; Korger P.; Funke B. *Melliand Textilver.* 1988, *69*, 286-287.
[34] Burtscher E.; Bechtold T.; Amann A.; Turcanu A.; Schramm C.; Bobleter O. *Melliand Textilber.* 1993, *74*, 903-907.
[35] Eriksson A.; Nyholm L. *Electrochim. Acta* 2001, *496*, 1113-1129.
[36] Eriksson A.; Nyholm L. *Electrochim. Acta* 1999, *44*, 4029-4040.
[37] Bechtold T.; Gutmann R.; Burtscher E.; Bobleter O. *Electrochim. Acta*, 1997, 42, 3483-3487
[38] Podsiadly R.; Sokolowska J.; Marcinek A.; Zielonka J.; Chrzescijanksa E.; Socha A. *Color. Technol.* 2003, *119*, 269-274.
[39] Malik W.U.; Gupta P.N. *Electroanal. Chem. and Interfac. Electrochem.* 1974, *54*, 417-426.
[40] Menek N.; Erdem E. *Chimica Acta Turcica* 1997, *25*, 155-157.
[41] Menek N.; Cakir O. *Portugaliae Electrochimica Acta* 1993, *11*, 175-183.
[42] Menek N.; *Analytical Letters* 1998, *31*, 275-282.

[43] Guaratini C.C.I.; Fogg A.G.; Zanoni M.V.B. *Electroanalysis* 2001, *13*, 1535-1543.
[44] Bechtold T.; Burtscher E.; Turcanu A. *Textile Research Journal,* 2001, *71*, 870-878.
[45] Dolezal J.; Langmyhr F.J. *Anal. Chim. Acta* 1972, *61*, 73-81.
[46] Marzinkowski J.M.; van Clewe B. *Melliand Textilberichte* 1998, *79*, 174-177.
[47] Schäfer T; Trauter J.; Janitza J. *Textilveredlung*, 1997, *32*, 79-83.
[48] Wehlmann U. *Melliand Textilberichte* 1997, *78*, 249-252.
[49] Majewska-Nowak K.; Winnicki T.; Wisniewski J. *Desalination* 1989, *71,* 127-135.
[50] Diaper C.; Correia V.M.; Judd S.J. *J. Soc. Dyers Color.* 1996, *112,* 273-281.
[51] Janitza J.; Koscielski S., *Int. Text. Bull.* 1996/4, 28-32.
[52] Levafix PN, Procion P, Procion PX livens up your reactive prints, Application brochure, 05.1032-00, DyStar, Frankfurt.
[53] Provost J.R., *J. Soc. Dyers Color.* 1992, *108,* 260-264.
[54] Brocks J., *textil praxis international*, 1992, *47,* 550-554.
[55] DyStar, Levafix®, Remazol® High Performance Reactive Dyes for all Requirements and Processes, 0.1010-00(E/D), 6/1999, DyStar, Frankfurt, Germany
[56] Bechtold T.; Turcanu A. *J. Appl. Electrochem.* 2004, *34,* 903-910.
[57] Bechtold T.; Burtscher E.; Turcanu A. *J. Chem. Technol. Biotechnol.* 2001, *76,* 303-311.
[58] Butnaru R.; Bertea A.; Dobrescu L. *Cell. Chem. Technol.* 2003, *37/3-4*, 477-486.
[59] Laschinger M. *Am. Dyest. Rep* 1999, 88, 23-27.
[60] Elgal G.M.; *Text. Chem. Color.* 1986, *18,* 15-20.
[61] Wilcock A.; Brewster M.; Tincher W. *Am. Dyest. Rep.* 1992, *81*,15-22.
[62] McClung S.M.; Lemley A.T. *Tex. Chem. Colour.* 1994, *26,* 17-22.
[63] Gutierrez M.C.; Pepio M.; Crespi M.; Mayor N. *Colour. Technol.* 2002, *117,* 356-361.
[64] Vilaseca M.M.; Gutierrez M.C.; Crespi M. *Melliand Textilber.* 2002, *83*, 558-560.
[65] Nevyas, J.; Lowy, A. *Trans. Am. Electrochem.* 1926, *50*, 12.
[66] Roessler, A.; Dossenbach, O.; Meyer, U.; Marte, W.; Rys, P. *Chimia* 2001, *55*, 879-882.
[67] Roessler, A.; Cretenand, D.; Dossenbach, O.; Meyer, U.; Marte, W.; Rys, P. *Electrochimica Acta* 2002, *47*, 1889-1995.
[68] Roessler, A.; Dossenbach, O.; Marte, W.; Rys, P. *J. Appl. Electrochem.* 2002, *32*, 647-651.
[69] Chamaut, H. *Bull. Soc. Internat. Electriciens* 1908, *8*, 13.
[70] Chamaut, H. *German Patent* 1907, *DE 223143.*
[71] Merk, C.; Botzem, J.; Huber, G:; Grund, N.; 2001 *Offenl. DE 19962155A1.*
[72] Merk, C.; Huber, G. ; Weiper-Idelmann, A. in The Electrochemical Society Inc., editor. Proc. 199th Meeting of the Electrochemical Society, Washington, DC, 2001, 121-124.
[73] Roessler, A.; Crettenand, D.; Dossenbach, O.; Rys, P. *J. Appl. Electrochem.* 2003, *33*, 901-908.
[74] Roessler, A.; Crettenand, D. *Dyes and Pigments* 2004, *63,* 29-37.
[75] Roessler, A. PhD-Thesis 2004, ETH Zürich.
[76] Roessler, A.; Dossenbach, O.; Rys, P. *J. Electrochem. Soc.* 2003, *150*, D1-D5.
[77] Roessler, A.; Dossenbach, O.; Marte, W.; Rys, P. *Dyes and Pigments* 2002, *54*, 141-146.
[78] Huber, G.; Weiper-Idelmann, A.; Kramer, A.; Golombek, R.; Frede, M.; Spiske, L.; Schehlmann, K.H.; Steuer, V. *European Patent* 2000, *EP 0808920.*

[79] Rengaranjan V.; Palanisamy R.; Sozhan G.; Narasimham K.C., *India. Bull. Electrochem.* 1993, *9(11-12),* 642-643.

[80] Tong C.; Tong W., Chinese Patent, 1988, CN 88-101765.

[81] Nishibe K. *Denki Kagaku oyobi Kogyo Butsuri Kagaku,* 1980 *48(4)* 267-270.

[82] Vijayaraghavan K.; Ramanujama T.K.; Balasubramanian N. *Color. Technol.* 2001 *117,*49-53.

[83] Varennes S.; Daneault C.; Levesque M. *Can. Appita J.* 1994 *47(1)*, 45-49.

[84] Fadali O. *Cellul. Chem. Technol.* 1991, *25(3-4),* 181-187.

[85] Nassar M.M. et.al., J. Appl. Electrochem. 1983 13(5) 663-667.

[86] Jakob H.; DelGrosso M.; Küver A. *Ger. Offen.* DE 198 43 571 (23.9.1998).

[87] Hampp N.; Ger. Offen. DE 197 23 889 (6.6.1997).

[88] Hu H.L.; Oloman C. *J. Paper Pulp Sci.* 1997, *23*, J232-J237

[89] Yamada N.; Yaguchi T.; Otsuka H.; Sudoh M. *J. Electrochem. Soc.* 1999, *146*, 2587-2591.

[90] M. Mickel M.; Kim H.; Noll S.; Hampp N. *J. Electrochem Soc.* 2003, *150*, E595-600.

[91] Gyenge E.; Oloman C.; Ames B. *Tappi. J.*, 1997, *80*, 194-202.

[92] Perng Y.S., Oloman C. *Tappi. J.* 1994, *77*, 115-125.

[93] T. Bechtold, P. Maier, W. Schrott, *Ger. Offen.* (2003) DE 10161265 (13.12.2001).

[94] Bechtold T.; Maier P.; Schrott W., *Colour. Technol.* in press.

[95] R.B. Chavan, B.C. Gaj, *Colourage* 1994, *41*, 21-25.

[96] Maier P.; Schrott W.; Bechtold T. *Melliand Textilberichte,* 2004, *85,* 880-884.

[97] Chong C.L.; Chu P.M. *Am. Dyest. Rep.* 1998, *87*, 13-19.

[98] Rengarajan V.; Palanisamy R.; Sozhan G.; Narasimham K.C. *Bull electrochem.* 1993, *9*, 642-643.

[99] Fadali O.A. *Egypt. Bull. Electrochem.* 1993, *9(1),* 21-24.

[100] Maier P.; Schrott W.; Bechtold T.; *Colouration Technology*, 121 (2005) 64-68.

[101] Morao A.; Lopes A.; Possoa de Amorim M.T.; Goncalves I.C. *Electrochim. Acta,* 2004 *49* 1587-1595.

[102] Fernandes A.; Morao A.; Magrinho M.; Lopes A.; Goncalves I.C. *Dyes and Pigments,* 2004, *61,* 287-296.

[103] H.J. Förster and W. Thiele, 7th Workshop Diamond Electrodes, Interlaken, Switzerland, 18.-20.10.2004.

[104] Hattori S.; Doe M.; Takahashi E.; Kurosu T.; Nara M.; Nakamatsu S.; Nishiki Y.; Furuta T.; Iida M. *J. Appl. Electrochem.,* 2003, 33, 85-91.

[105] Ceron-Rivera M.; Davila-Jimenez M.M.; Elizalde-Gonzalez M.P. *Chemosphere,* 2004, *55*, 1-10.

[106] Van Hege K.; Verhaege M.; Verstraete W. *Electrochem. Commun.,* 2002, *4,* 296-300.

[107] Chen X.; Chen G.; Yeu P.L. *Chem. Eng. Sci.,* 2003, *58,* 995-1001.

[108] Shen Z.; Wang W.; Jia J.; Ye J.; Feng X.; Peng A. *J. Hazard. Mat.,* 2001, *B84,* 107-116.

[109] Sanroman M.A.; Pazos M.; Ricart M.T.; Camselle C. *Chemosphere*, 2004, *57,* 233-239.

[110] Marselli B.; Garcia-Gomez J.; Michaud P.A.; Rodrigo M.A.; Comninellis C.J. *J. Electrochem. Soc.* 2003, *150*, D79-D83.

[111] Horne C.M.; *Text. Chem. Color.* 1995, *27*, 27-32.

[112] Baumgarte U., *Melliand Textilber.* 1987, *68*, 276-280.

[113] Etters J. N.; *Text. Chem. Color.* 1995, *27*, 17-22.

[114] Aspland J.R.; *Text. Chem. Color.* 1992, *24/3* () 21-24.

[115] Aspland J.R.; *Text. Chem. Color.* 1992, *24/4* 27-31.

[116] Bechtold T.; Burtscher E.; Gmeiner D.; Bobleter O. *J. Electroanal. Chem. Interfac. Electrochem.* 1991, *306,* 169-183

[117] Bechtold T.; Burtscher E.; Amann A.; Bobleter O. *Angew. Chemie* 1992, *104/8,* 1046-1047; *Angew. Chem. Int. Ed. Engl.* 1992, *31/8,* 1068-1069.

[118] Bechtold T.; Burtscher E.; Amann A.; Bobleter O. *J. Chem. Soc. Faraday Trans.* 1993, *89,* 2451-2456.

[119] Bechtold T.; Burtscher E.; Bobleter O. *Research Trends, Current Topics in Electrochemistry,* 1998, *6,* 97-110.

[120] Mohr S.; Bechtold T. *J. Appl. Electrochem.* 2001, *31,* 363-368.

[121] Bechtold T.; Burtscher E.; Turcanu A. *J. Electroanal. Chem.* 1999, *465,* 80-87.

[122] Bechtold T.; Berktold F.; Turcanu A. *J. Soc. Dyers Color.* 2000 *116,* 215-221.

[123] Bechtold T.; Turcanu A.; Burtscher E.; Bobleter O. *Textilveredlung,* 1997, *32,* 204-209.

[124] Bechtold T.; Burtscher E.; Kühnel G.; Bobleter O. *Journal of the Society of Dyers and Colorists* 1997, *113/4,* 135-144.

[125] Bechtold T.; Burtscher E.; Turcanu A.; Bobleter O. *Text. Res. J.* 1997, *67*, 635-642.

[126] Bechtold T.; Turcanu A. *J. Electrochem. Soc.,* 2002, *149,* D7-D14.

[127] Bechtold T.; Burtscher E.; Turcanu A. *J. Chem. Soc. Dalton Trans.* 2002, 2683-2688.

[128] Bechtold T., Turcanu A., *J. Appl. Electrochemn.* 2004, *34,* 1221-1227.

[129] Schrott W.; Bechtold T. *Textilveredlung,* 2003, *38,* 19-22.

[130] Blatt W.; Schneider L. *Melliand Textilber.* 1999, *80*, 624-626.

[131] Schmidt, V.M. *"Elektrochemische Verfahrenstechnik"*, 2003, VCH-Wiley, Weinheim.

[132] Cœuret, F. *„Ingénierie des procédés électrochimiques*, 2003, Ellipses Édition, Paris.

[133] Bechtold T., Burtscher E., Bobleter O., Blatt W., Schneider L., *Chem. Eng. Technol.* 1998, *21,* 877-880.

[134] Bechtold T., Turcanu A., *J. Electrochem. Soc.* 2002, *149,* D7-D14.

[135] Frede M.; Roske E.; Pütter H. Ger. Offen. De 195 13 839 (12.4.1995)

[136] Vallières C.; Matlosz M. *J. Electrochem. Soc.* 1999, *146*, 2933-2939

[137] Hinman A.S.; Wiebe P. *Anal. Chem.* 1995, *67*, 694-700.

[138] Bechtold T., Burtscher E., Turcanu A., Bobleter O., *J. Electrochem. Soc.* ***143**/8* (1996) 2411-2416.

[139] Brunner, H.; Bechtold, T.; Thiele, W.; Förster, K.H.; Schrott, W. *German Patent Appl.* 2004, *DE 102004023161.3.*

In: New Developments in Electrochemistry Research
Editor: Magdalena Nunez, pp. 55-83
ISBN: 1-59454-544-8

Chapter 2

Kinetic Aspects of Electrical Communication in Amperometric Biosensors

Victor Roşca[1] and Ionel Cătălin Popescu[*2]

[1] Schuit Institute of Catalysis, Laboratory of Inorganic Chemistry and Catalysis, Eindhoven University of Technology, Eindhoven, 5600 MB, The Netherlands

[2] Department of Physical Chemistry, Faculty of Chemistry and Chemical Engineering, "Babeş-Bolyai" University, RO-400084, Cluj-Napoca, Romania

Abstract

Amperometric biosensors form one of the most prominent classes of analytical devices based on biological recognition. Next to providing a natural way for the signal transduction in the redox enzymes–based sensors, amperometric transduction offers remarkable selectivity, sensibility, ultimately resulting in low detection limits and a low cost of the final device. At the same time, similarly to the great majority of the analytical devices based on biological recognition, the biocatalyst (enzyme) causes major problems in building and exploitation of an amperometric biosensor. Accordingly, beside the stability of the biological element, electrical communication between the signal transducer and the biocatalytic cycle of the enzyme is a matter of primary concern. Therefore, a thorough understanding of the mechanism and kinetics of the signal generation in amperometric biosensors is a prerequisite for a rational and efficient design of amperometric biosensors.

In this contribution, we address recent advances in the amperometric biosensor research, focusing specifically on the kinetic aspects of the electrical communication between the enzyme redox center and the electrode. Accordingly, different transduction schemes are considered, relying mainly on the data gained from mechanistic and kinetic studies.

In the first section, the reader is introduced into the field of amperometric biosensors, and the topic and the goals of this contribution are stated. The next two sections are dedicated to the amperometric biosensors based on two main groups of oxidoreductases: dehydrogenases and peroxidases. The electrical communication with the catalytic cycle

* Prof. Dr. I.C. POPESCU, Tel. +40-264-59-58-72, Fax. +40-264-59-08-18; cpopescu@chem.ubbcluj.ro

of a dehydrogenase is in most cases carried out through the electrochemical regeneration of the active form of the co-factor. For the case of peroxidases, the direct electron transfer to the enzyme active center is perceived as the most attractive way of the enzyme reactivation and at the same time signal transduction.

1. Introduction

Exceptional selectivity and highly specific action of some biochemical systems make their use with analytical purposes very attractive and, therefore, resulted in emerging of a new prominent class of analytical devices referred to as biological sensors or biosensors [1,2]. In biosensors design the specific molecular recognition of a selected biochemical system is used for generating a physical/chemical (concentration-dependent) signal. Accordingly, a biosensor may be defined as a compact analytical device incorporating a biological recognition element closely associated with a suitable signal transducer [3].

Biological recognition element (BRE) is the key constituent of a biosensor, since its action determines the maximum selectivity and specificity of the final analytical device. There are four major types of biosensors: (bio)affinity biosensors, (bio)catalytic biosensors, trans-membrane biosensors and cell sensors [4]. At the same time, with all diversity of BRE employed so far, the function of all biosensors is based on two fundamental types of recognition: (i) one implies selective binding associated with a catalytic process (e.g. recognition in the system enzyme-substrate) and (ii) another – specific intermolecular interactions, or affinity interactions (e.g. recognition in the system antibody-antigen or receptor-hormone). Accordingly, most of the trans-membrane biosensors are based on bioaffinity as well as on functioning of some biocatalytic subsystems [5].

The choice of the signal-transducing scheme is largely determined by the nature of the effect, which accompanies the biological recognition. Electrochemical, optical, piezoelectric and calorimetric transducers are the most important types of transducers used for biosensor design [6]. An effective and, at the same time, efficient transduction scheme would require achievement of an optimal physical and functional contact between the biological recognition element and transducer, and commonly requires presence (immobilization) of the BRE in close proximity (or in direct contact) with the surface, within a well-organized thin (μm) layer [7]. An efficient immobilization technique should not affect decisively the efficiency of the biological recognition act(s). Accordingly, the biological molecule should keep its functional identity, its active center should not be physically blocked and the transport phenomena should be facile. An BRE may be immobilized by physical or chemical (covalent bonding) adsorption on the surface, retained behind a membrane, entrapped physically or/and chemically into a polymer matrix [5]. At the same time, complexity of biochemical systems and the challenge of an effective signal transduction impose necessity of a controlled (microscopic level) design of the interface. Electrostatically-controlled adsorption of charged species, such as bio- and synthetic polymers, covalent or bioaffinity coupling to ordered molecular monolayers (Langmuir-Blodgett or self-assembled monolayers), embedding in a lipid membrane and other techniques, in combination with available miniaturization tools would allow controlled design of interfaces with high level of structural and functional order [4,8-21].

Recent evaluations of the transduction schemes for biosensors design [22] indicate that amperometric detection is one of the most attractive schemes, particularly in regard to its (relatively) simple construction, remarkable sensitivity and low cost of the final device. Thus, a successful amperometric transduction would allow lowering the detection limit up to 10^{-9} M, while the concentration region of a linear response would be three to four decades [1]. Redox enzymes are commonly employed in the position of the biological recognition element in amperometric biosensors, since their catalytic cycle implies the electron transfer processes. Function of an oxidoreductase is to catalyze transformation of the target molecule resulting in formation or consumption of an electrochemically active species. The process of transduction is accomplished by applying a constant potential to the working electrode and monitoring the resulted current as a function of time. Importantly, a measurable current will be generated only if there is a way to regenerate the active center (or natural redox partner) of the enzyme and, thus, perpetuate the recognition/conversion process [5].

Enzymes are highly specific and highly active natural catalysts for chemical reactions, which are fundamentally important for a normal function of living cells [23]. Enzymes are complex chemical compounds with molecular weight between 10^4-10^6 g mol^{-1}. Although, enzymes are mainly proteins, their catalytic activity commonly implies a non-proteinic component with a low molecular weight, called cofactor. The latter can be an organic molecule or a metal complex. If the cofactor is loosely bound to the proteinic chain (apoenzyme) of the enzyme, then the cofactor is called coenzyme. Otherwise, the cofactor forms the so-called prosthetic group.

With all complexity of the enzyme action, enzymatic catalysis essentially follows the laws of conventional chemical reactions [24]. Accordingly, the following factors are perceived responsible for catalytic activity of enzymes [25]: (1) high affinity between enzyme and substrate, which increases the probability of formation of the enzyme-substrate complex (proximity effect); (2) well-defined spatial orientation of the reagents (orientation effect); (3) impact of neighboring reactive groups (polyfunctional catalysis includes acid-base catalysis, “pull-push”-type mechanisms etc); (4) the activation of the substrate by redistribution of the local electronic density caused by an active electronegative center (polarization effect). A combined action of these factors brings about a dramatic increase of the overall rate of the process, since, formally, a reaction of a higher kinetic order (two or three) is effectively replaced by a reaction of the first order. At the same time, the complexity of the enzyme action means that the enzyme activity may be quite susceptible to external factors, the issue to be considered when using enzymes with analytical or any other purpose. Thus, enzymes are commonly efficient in a quite narrow region of pH. Immobilization may cause conformational changes in the protein chain and, thus, affect the enzyme activity. Then, a complex chemical environment may cause various deactivation processes.

As a function of the nature of the chemical reaction catalyzed, enzymes are divided in six main classes [23]. Among these, oxidoreductases (oxidation-reduction), hydrolases (hydrolysis reactions) and lyases (addition or removal of groups to form double bond) are potentially important for use in amperometric biosensors; the first class mentioned being the most widely employed and studied.

Electrochemical recycling of the enzyme prosthetic group (e.g., heme, pyrroloquinoline quinone (PQQ), flavin adenine dinucleotide (FAD) etc.) at the electrode surface would provide a straightforward way of the signal transduction in amperometric biosensors. However, the kinetics of the direct electron transfer between the electrode surface and the prosthetic group, which is hidden inside the protein globule, is sluggish because of high reorganization energy and significant electron transfer distances, as predicted by the Marcus theory [26,27]. For the case of cofactor-dependent enzymes, it would be necessary to (electrochemically) recycle the active form of the cofactor (e.g. nicotinamide adenine dinucleotide (phosphate) $NAD(P)^+$, flavin adenine mononucleotide (FMN) etc.), which is not a trivial task at all. Therefore, mechanism and kinetics of communication between electrodes and biocatalytic redox systems is a central issue in fundamental and applied biosensor research. Accordingly, the advances in the field of amperometric biosensor research may be presented in terms of "generations" of biosensors; the distinction between them being based on the mode the electrical communication with the transducer is achieved.

In the "*first generation*" amperometric biosensors (Figure 1, scheme A), an indirect electrical coupling between the enzyme catalytic cycle and the electrode is achieved via electrochemical transformation of a freely diffusing natural redox couple, which is involved in the enzymatic reaction as reagent (e.g. O_2), product (e.g. H_2O_2) or natural redox co-substrate (e.g. FAD or $NAD(P)^+$). Such a signal transduction usually requires applying of high electrode potentials, thus inevitably causing the interfering electrode processes. Furthermore, since the electrochemically active species diffuse freely in solution, the mass transport has a strong influence on the biosensor response. For example, insufficient partial pressure of oxygen in blood is a serious problem for "*in-vivo*" applications of glucose oxidase-based biosensors.

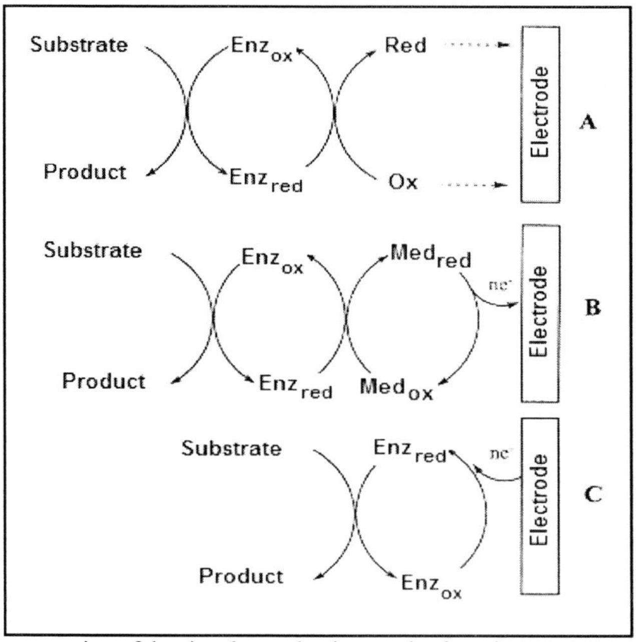

Figure 1. Schematic representation of the signal transduction mechanisms in amperometric biosensors. See also the text for details.

In order to overcome all difficulties related to the "*first-generation*" amperometric biosensors, in the "*second generation*" amperometric biosensors (Figure 1, scheme B) artificial mediators are employed to effectively recycle the cofactor or the enzyme active center. Fundamental requirements toward a successful redox mediator are [5]: (i) the mediator redox couple should be a couple with a well-defined stoichiometry; (ii) the mediator should be sufficiently soluble and stable under conditions of its exploitation; (iii) the mediator formal potential should be adequate to the enzymatic process; (iv) a successful redox couple playing the mediator role is characterized by a fast heterogeneous electron exchange with the electrode and fast homogeneous electron exchange with the enzyme redox center or, alternatively, enzyme cofactor. In principle, choosing a suitable artificial mediator proved to be a fruitful strategy for overcoming drawbacks related to using the enzyme natural partners. Thus, by using an appropriate redox mediator (freely diffusing or immobilized on the electrode surface), the overpotential of the electrochemical oxidation of NADH coenzyme may be reduced significantly, with a high selectivity towards production of the enzymatically active NAD^+ [28]. An important prerequisite for an effective use of a redox mediator is the mobility of the mediator, since this would be important for reaching the distances and orientation, which would allow the electron transfer to take place. From this point of view, a freely diffusing mediator would be the simplest solution for assuring the mediator mobility. However, this would mean the reagent leakage into the solution out of the interfacial region, thus seriously affecting the performance of the biosensor. To avoid this, the redox mediator is usually immobilized in the interfacial region. Importantly, the immobilization technique must assure sufficient mobility and a sufficiently high (local) concentration of mediator in the interfacial region. For example, the mediator redox centers may be immobilized on the electrode surface or, alternatively, bound to the enzyme surface via long flexible chains [16]. Another, in many respects attractive, technique would be entrapment of the enzyme in a polymer matrix of a "redox hydrogel" or a conducting polymer [16,29].

The last remark leads us to discussion of the "*third generation*" amperometric biosensors (Figure 1, scheme C), i.e. devices in which the enzyme is obligatorily an integrated part of the sensor: the biocatalyst is immobilized on the electrode surface or entrapped in a polymer matrix. These immobilization techniques would imply possibility of either a direct electron transfer between the enzyme prosthetic group and the electrode or the possibility of a mediated electron transfer between the enzyme prosthetic group and the modified electrode surface via a mediator, which does not diffuse freely (e.g. an enzyme entrapped in a polymer matrix of a redox polymer). An important consequence of the enzyme (and mediator) physical integration within the transducer interface is the reagentless character of the resulting biosensor. Note that biosensors based on the enzymes entrapped in the redox polymers presents characteristics of both the "second" and "third generation" biosensors [16]. Indeed, the redox mediator is responsible for the charge transport between the enzyme redox center and the electrode surface. At the same time, mediator is not involved in free diffusion. Therefore, the polymer matrix may be, at least formally, viewed as a part of the electrode, and, therefore, respect the criteria of the "*third generation*" biosensors.

With all difficulties of accomplishing, a direct electron transfer between the active center of the enzyme and electrode surface, avoiding any mediation, it is perhaps the most attractive way of electrical communication in amperometric biosensors. Firstly, absence of a mediation step would mean better selectivity and a simpler design. Secondly, the performance of the biosensor may be improved by manipulating the enzyme properties, for instance, by applying modern techniques of molecular engineering and/or the electrode surface design at a molecular level [9,15,30,31]. Only the enzyme molecules, which are in direct contact with the electrode surface, may exchange electrons with the electrode, provided the orientation is adequate and the tunneling distance is achieved [29]. In this regard, an anisotropic (oriented) immobilization, e.g. on the surface of a self-assembled monolayer, presents an attractive way for optimization of the direct electron transfer [29,32-34]. Direct electron transfer is characteristic to a quite limited number of redox enzymes [35]. Among these, peroxidases are most important for application in amperometric biosensors [36].

The above, even a brief, incursion into the issue of the electron transfer mechanisms in amperometric biosensors clearly indicates the fundamental importance of mechanistic and kinetic investigations of the signal transduction. Indeed, a well-defined electron transfer pathway within a structurally and functionally well-defined electrode-electrolyte interface would be the most important prerequisite for building a successful biosensor [37]. Next, extensive kinetic studies would naturally succeed the mechanistic studies, since kinetic analysis of the system would allow (i) testing the performance of the transduction scheme and (ii) providing reliable criteria for the biosensor optimization. Furthermore, the immobilization or, generally, the electrode modification technique(s) should be scrutinized in order to improve the efficiency of the electrical communication within the electrode architecture.

In this short review-type contribution, we shall cover the issue of the electrical communication between the enzyme active centers and the electrode surface, aiming at emphasizing the role of mechanistic and kinetic studies at early stages of the biosensor design. In the subsequent sections, we shall consider the amperometric biosensors based on two groups of enzymes, dehydrogenases and peroxidases. The electrical communication with the catalytic cycle of dehydrogenases is in most cases carried out through the electrochemical regeneration of the active form of the cofactor. For the case of peroxidases, the direct electron transfer to the enzyme active center is perceived as the most attractive way for the enzyme regeneration and, at the same time, for the signal transduction. Therefore, covering the electrical communication for these two groups of enzymes with electrodes would mean covering two fundamentally important cases in modern biosensor research.

2. Dehydrogenase-Based Amperometric Biosensors

Dehydrogenases form a quite numerous class of redox enzymes capable of a reversible electron (hydrogen) transfer from a substrate to an acceptor, but not to molecular oxygen. Dehydrogenases may be classified by the strength of bonding between the apoenzyme and cofactor. FMN, FAD and PQQ are all strongly bound to the apoenzyme, while the nicotinamide coenzymes (NAD^+ and $NADP^+$) are bound to the apoenzyme during the reaction.

The catalytic activity of dehydrogenases, which are dependent on the pyridinenucleotides, may be represented schematically as follows:

$$SH_2 + NAD(P)^+ + \text{apoenzyme} \rightarrow S + NAD(P)H + \text{apoenzyme} \quad (1)$$

Essentially, this type of transformation represents a net transfer of a hydride ion (H^-) from a substrate to the cofactor. However, at present it is not clear yet whether the reaction involves the hydride transfer in a single step or follows a stepwise radical mechanism [38]. The reaction pathway typically belongs to the two-substrate-type mechanism. First, NAD^+ forms a complex with the apoenzyme and, subsequently, this complex (holoenzyme) binds the substrate. Then, desorption of reaction product occurs, followed by the dissociation of the apoenzyme-cofactor complex. Note that some dehydrogenases are capable of employing both NAD^+ and $NAD(P)^+$ in their reactions, while others are specific to one of the cofactors.

Construction of the amperometric biosensors based on dehydrogenases, which uses nicotinamide coenzymes, is quite a challenge, particularly in relation to the necessity of immobilization of both the apoenzyme and the coenzyme [28,39]. The electrical transduction through NADH oxidation presents a number of decisive advantages for the amperometric detection. Thus, besides providing the electrical communication with the enzyme catalytic cycle, NADH oxidation is used for recycling the enzymatically active NAD^+, which would assure the continuity of the biocatalytic process. Furthermore, "*in-situ*" recycling of NAD^+ would sensibly decrease the final cost of the analytical device.

The remaining part of this section will be dedicated to the issue of electrochemical (electrocatalytic) oxidation of NADH, i.e. one of the central issues in the amperometric biosensors research. The nicotinamide cofactors have similar structure (Figure 2) and have unique redox characteristics. The natural redox cycle of the nicotinamide cofactors implies a transfer of two electrons and one proton and may be (at least formally) viewed as a hydride transfer to the forth carbon in the nicotinamide nucleus:

$$NAD(P)^+ + H^+ + 2e^- \leftrightarrows NAD(P)H \quad (2)$$

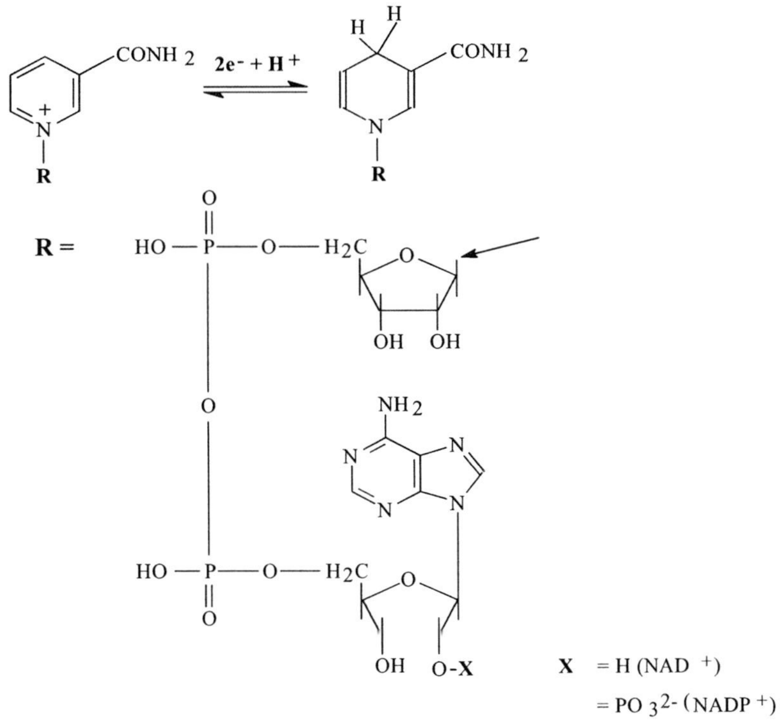

Figure 2. The structure and reduction-oxidation reactions of NAD(P)H.

The formal standard potentials of the NAD^+/NADH and $NADP^+$/NADPH couples are -0.561 V and -0.565 V vs. saturated calomel electrode (SCE), respectively (pH 7) [40]. At the conventional electrode materials, the voltammetric response corresponding to the cofactor oxidation is observed with overpotentials between ca. 1 and 1.3 V (pH 7). Despite the complexity of the NADH electrochemical oxidation at conventional electrode materials, recycling of the enzymatically active NAD^+ seems to be possible [41,42]. At the same time, poisoning of the electrode surface as a result of various side reactions, ultimately resulting in low efficiency of NAD^+ production and a moderate rate of the reaction itself, makes the conventional electrode materials (e.g., platinum, gold, various types of graphite) difficult to employ in the design of amperometric biosensors incorporating dehydrogenases.

With some disagreement on the nature of the elementary steps and the rate-determining step of the reaction, the direct electrochemical reduction of NADH is commonly described in terms of an ECE mechanism [43-50]:

$$NADH \leftrightarrows NADH^{\bullet+} + e^- \quad (3)$$
$$NADH^{\bullet+} (+ B) \leftrightarrows NAD^{\bullet} + (B)H^+ \quad (4)$$
$$NAD^{\bullet} \leftrightarrows NAD^+ + e^- \quad (5)$$

where B is a base. High overpotentials for NADH oxidation may originate in a high potential of the $NADH^{\bullet+}$/ NADH redox couple, 0.78-0.81 V vs. SCE [51-53]. Provided that the first electron transfer is indeed the rate-determining step; then, assuming that deprotonation of the $NADH^{\bullet+}$ cation-radical is sufficiently fast (6×10^6 s^{-1}, according to some estimations) [52-54] and, finally, taking into consideration the acidity of the $NADH^{\bullet+}$ cation-radical ($pK_a\approx-4$) [55], an ECE mechanism seems applicable to the direct electrochemical oxidation of NADH. Note, however, that there is still some controversy on the rate-determining step of the process. Thus, there is some evidence pointing towards the proton transfer step as the rate-determining step [49,56,57]. The adsorption/desorption equilibrium, role of the oxygen-containing species and side reactions, particularly those involving $NAD^{\bullet}$ radical, also contributes to the complexity of the process.

A classical in the biosensor research approach is using of redox mediators, which are able to sustain an effective and efficient "shuttling" of electrons between the cofactor and the electrode surface. From the above discussion about the direct electrochemical oxidation of NADH it is clear that a successful mediator for NADH oxidation should be able (i) to decrease the overvoltage of NADH oxidation and (ii) to assure a fast and efficient (no side reactions) recycling of enzymatically active NAD^+. More specifically, these requirements would mean the ability to assure a fast transfer of the first electron and the proton, combined with the ability to form a charge-transfer complex and thus avoid side reactions of $NAD^{\bullet}$. This is why, the ability of some one-electron mediators to catalyze (mediated electrocatalysis) the NADH oxidation seems to be quite surprising.

Miller et al. [51,52] carried out a detailed study of the mechanism and kinetics of NADH oxidation by a series of the ferrocene derivatives in solution. Since the mediator is able to exchange one electron only, the proton is donated to the solvent or, more generally, to a base B. Therefore, the NADH oxidation by one-electron mediators mimics the direct electrochemical oxidation of NADH at the electrode surfaces. The homogeneous oxidation of NADH by ferrocenes starts with the electrochemical generation of the active form of mediator:

$$Fc \leftrightarrows Fc^+ + e^- \quad (6)$$

The oxidation reaction itself follows an ECE mechanism:

$$NADH + Fc^{+} \underset{k_{-1}}{\overset{k_1}{\Leftrightarrow}} NADH^{\bullet +} + Fc \tag{7}$$

$$NADH^{\bullet +} + B \underset{k_{-2}}{\overset{k_2}{\Leftrightarrow}} NAD^{\bullet} + BH^{+} \tag{8}$$

$$NAD^{\bullet} + Fc^{+} \underset{k_{-3}}{\overset{k_3}{\Leftrightarrow}} NAD^{+} + Fc \tag{9}$$

The authors showed that the reaction rate is the first order in respect to both [Fc^{+}] and [NADH] (second order globally) and is independent of the pH value. Furthermore, addition of either NAD^{+} or Fc proved to have no effect on the reaction rate. Two equivalents of Fc^{+} react with one equivalent of NADH, both reactants being consumed completely. The yield in the enzymatically active NAD^{+} was up to 75%. Based on this evidence, the authors concluded that similarly to direct electrochemical reduction, the first electron transfer is the rate-determining step (rds), provided the product k_2[B] is higher than k_1[Fc].

Importantly, the first electron transfer is perceived as a complex process consisting of three steps: (1) the approach of the reactant species and formation of a complex, (2) the intra-complex electron transfer and (3) the dissociation of the resulted complex and the away-diffusion of the products. The actual rate- limiting step would depend on the formal potential of the ferrocene: for ferrocenes with a low formal standard potential, the separation of the Fc-$NADH^{\bullet +}$ couple would be the rate determining step, while for those with more positive formal standard potentials, the electron transfer itself is the rate limiting. Note that the reaction is irreversible, since (i) Fc species are consumed at the electrode and (ii) the redox potential of the couple $NAD^{\bullet}$/ NAD^{+} is at least 1.3 V more negative than that of the Fc^{+}/Fc couple [58]. As a final remark, despite the fact that one-electron mediators seem of little practical value for dehydrogenase-based biosensors, studies on these mediators were certainly insightful for a better understanding of the mechanism of the NADH electrochemical oxidation.

Early studies on the electrochemical oxidation of NADH at carbon electrodes pointed to the possibility of electrocatalytic activity of some surface species toward NADH oxidation. Thus, formation and the catalytic activity of oxygen-containing species, particularly quinoid derivatives, were held responsible for a decrease of the overvoltage for NADH electrochemical oxidation [59]. Subsequent testing of *o*- and *p*-quinones, followed by *o*- and *p*-diimines and a series of the so-called redox dyes (phenoxazines, phenothiazines, phenazines), demonstrated the ability of these 2e/(1-2)H^{+} mediators to effectively and efficiently convert NADH into enzymatically active NAD^{+} [60-62].

In several combined electrochemical and spectroscopic studies, Miller et al. provided quite a detailed picture of NADH oxidation by a series of quinones and quinonimines [63-65]. Significantly, the authors provided clear evidence for an hydrogen transfer from the position 4 of the nicotinamide ring of NADH to the quinone oxygen and an 1:1 stoichiometry of the process. The reaction proceeds trough formation of a charge transfer complex between NADH and the quinone (Q). The net hydride transfer takes place most probably stepwise:

$$NADH + Q \leftrightarrows [NADH\ Q] \xrightarrow{rds} [NAD^{\bullet}QH^{\bullet}] \quad (10)$$
$$[NAD^{\bullet}QH^{\bullet}] \rightarrow [NAD^{+}QH^{-}] \quad (11)$$
$$[NAD^{+}QH^{-}] \rightarrow NAD^{+} + QH^{-} \quad (12)$$

An alternative scenario implies the first electron transfer to be the rate-limiting step, followed by fast hydrogen transfer and, finally, dissociation of the complex. Interestingly, Miller and coworkers also suggested the possibility of a hydride transfer in a single step. Although this scenario cannot be ruled out, the stepwise mechanism would provide more flexibility in explaining the experimental data for NADH electrocatalytic oxidation by mediators in solution or immobilized mediators, including pH effect and side reactions [60,66,67]. Diimines, known to be even stronger hydride acceptors, show a behavior quite similar to that of the quinines [63,64]. Importantly, these studies showed that the superiority of the two-electron (proton) mediators (over the one-electron mediators) is related to their ability to form a charge transfer complex and, therefore, to stabilize the reacting NADH molecule against possible side reactions [61].

The model kinetic studies performed in solution were followed by numerous studies on mediated electrocatalysis of NADH oxidation by immobilized $2e^{-}/(1\text{-}2)H^{+}$ mediators. Several comprehensive and detailed reviews on the subject are available [28,39,60,61,68]. It is important to realize that when an application to biosensor technology is aimed at, the requirements toward a suitable mediator are quite strict. Next to the major objectives, i.e. to decrease the overvoltage, high values for the observed second-order rate constant and a well-defined stoichiometry of the reaction with NADH, the chemical stability of the mediator and the stability of the transducing interface under exploitation conditions become equally important [39,68]. Therefore, selection or a rational design of the mediators for NADH electrocatalytic oxidation would imply testing the mediator activity and selectivity, and, ultimately, its adaptability to the modification procedures employed. Physical adsorption, covalent binding to the surface, incorporating into the carbon paste or into a polymer matrix are most frequently employed immobilization techniques. All these techniques may affect the kinetic parameters of the mediator.

As already mentioned in this section, a successful mediator should be able to effectively decrease the overpotential of the NADH electro-oxidation, which would require mediators with the formal potential as low as possible (ideally around 0 V vs. SCE, pH 7). At the same time, the electron transfer between the electrode surface and mediator should be fast, at least sufficiently fast to keep up with the rate of NADH oxidation by the oxidized mediator. Typical values of the heterogeneous rate constant (k_s) for redox dyes are between 1 and 50 s^{-1} [60,68], higher values, up to 600 s^{-1}, were reported for some nitroderivatives [69,70]. Commonly, this heterogeneous rate constant is determined from the cyclic voltammetry experiments, which actually gives the lowest values of the constant [71,72]. The second order rate constant of the reaction between the mediator and NADH may be evaluated from the potential sweep measurements [73] or, more adequately, by using the rotating disc electrode technique [74]. The Koutecky-Levich formalism combined with the Michaelis-Menten-type mechanism for NADH oxidation allow estimation of the observed second-order rate constant

(k_{obs}), provided the kinetically-controlled regime (transport is sufficiently fast) is achieved [60,68,74]. The rate constant k_{obs} ranges from 10^0-10^2 for one-electron mediators to 10^3-10^6 M^{-1} s^{-1} for two-electron/proton mediators [60-62,68,75,76], while the desired values are between 10^6 and 10^7 M^{-1} s^{-1}. Note that k_{obs} shows variation with the pH, lower values being observed for higher pH values. The effect of the pH is difficult to study in detail, because the electrochemistry of both reactants (the mediator and NADH) is susceptible to the changes of pH [60,68].

We shall continue with presenting an interesting case of mediators for NADH electro-oxidation, namely the case of dibenzo-dithia-diazapentacene (DDD) derivatives (Figure 3) [77,78]. The DDD derivatives are one-electron/one-proton mediators and, therefore, an examination of their electrocatalytic activity towards NADH oxidation is certainly interesting from a fundamental point of view. Following the idea of increasing chemical and electrochemical stability of redox mediators by extension of the intramolecular conjugation system [60], the use of these new phenothiazine derivatives, which have two linearly condensed phenothiazine moieties, was proposed [79]. Later, this strategy was checked on four newly synthesized dibenzodithiadiazapentacene derivatives, immobilized by adsorption on different electrode materials in order to obtain modified electrodes for NADH electrocatalytic oxidation [80,81]. The DDD derivatives are poorly soluble in water and, besides, contain moieties, which may promote interaction with surface (extended aromatic rings, sulfur atoms). Therefore, these compounds were successfully immobilized and studied on spectrographic graphite [79,81], graphite powder - zirconium phosphate (carbon) paste [81] and gold [80] electrodes. In aqueous solutions, the DDD derivatives are able to transfer reversibly only one proton and one electron (Figure 4); subsequent electron/proton transfer causes an irreversible decomposition of the molecule. The formal standard potential of the DDD derivatives shows a shift of ca. 60 mV per pH unit (except when immobilized with the zirconium phosphate [81]). All compounds are remarkably stable in a wide range of pH (from 2 to 12) [79,80].

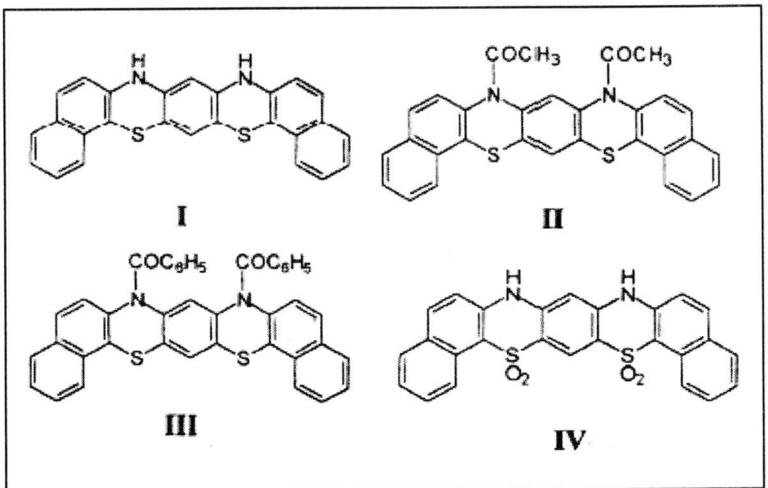

Figure 3. Structure of the dibenzodiazadithiapentacene (DDD) derivatives. I - 16*H*,18*H*-dibenzo[c,l]-7,9-dithia-16,18-diazapentacene (DDDH); II - dibenzo[c,l]-16,18-diacetyl-7,9-dithia-16,18-diazapentacene; III - dibenzo[c,l]-16,18-dibenzoyl-7,9-dithia-16,18-diazapentacene; IV - 16*H*,18*H*-dibenzo[c,l]-7,9-dithia-16,18-diazapentacene-7,7,9,9-bis-dioxide.

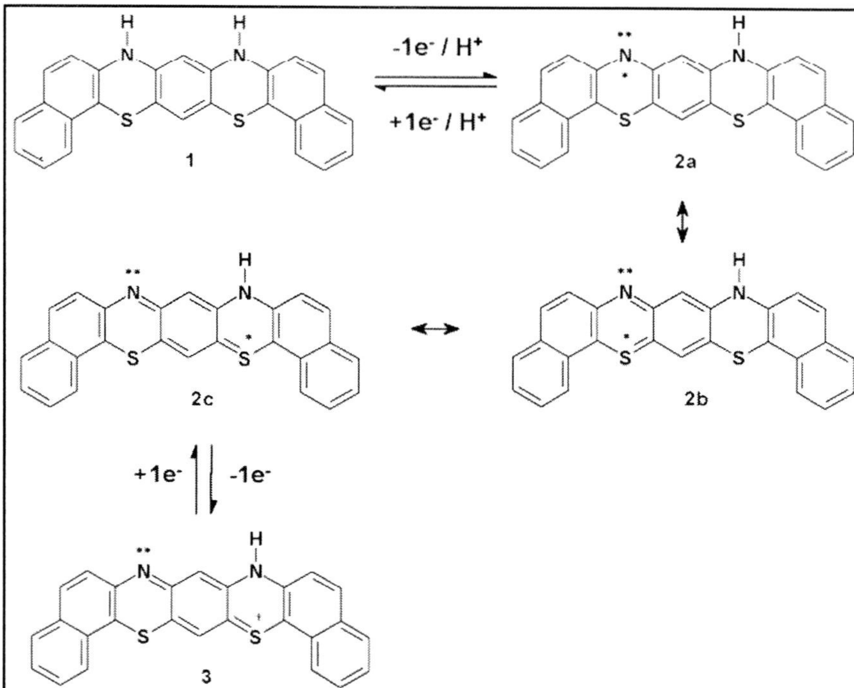

Figure 4. The reaction scheme for the DDDH reduction-oxidation. 2a-c are mesomeric structures for the DDD• radical.

The DDD derivatives are characterized by a relatively fast electron transfer kinetics, when immobilized on spectrographic graphite or carbon paste (k_S = 20-135 s^{-1}) [79,81], and a significantly lower when immobilized in carbon paste with zirconium phosphate (ca. 1 s^{-1}) [81]. Interestingly, the same compound, DDDH, shows different electron transfer rate as a function of the electrode material: ca. 1 s^{-1} on gold [80], ca. 30 s^{-1} on spectrographic graphite [79] and ca. 60 s^{-1} when entrapped into a carbon paste [81] (all at pH 7, Table 1). This difference can be tentatively attributed to the different surface orientation of the molecule at the surface and, therefore, different electron transfer distances. Thus, it was postulated that on gold, DDDH binds to the surface through sulfur atoms, the molecule "plane" being nonparallel to the surface. This would mean that the active center(s) of the molecule (assumedly, one of the two nitrogen atoms) is distant from the surface. On graphite, the molecule is most probably oriented parallel to the surface, due to the strong interactions between the aromatic moieties of the molecule with the graphite planes, thus placing the nitrogen atoms close to the surface [79] and ultimately resulting in shorter electron transfer distances. Even higher rates of the heterogeneous electron transfer for DDDH immobilized in carbon paste may be related to even closer proximity of the molecule active center to the electrode "surface". The DDDH adsorption on the carbon (graphite) - zirconium phosphate paste makes the mediator formal potential virtually independent of the pH. At the same time, the heterogeneous rate constant of the mediator drops to ca. 1 s^{-1} [81].

Table 1. Electrochemical and electrocatalytic parameters of selected DDDH-modified electrodes

Electrode	pH	E°' (V vs. Ag | $AgCl/KCl_{sat}$)	k_s (s^{-1})	$k_{obs\,[NADH]=0}$ (10^2 M^{-1} s^{-1})
SG[a] / DDDH	6	0.38	34	10.5
	7	0.32	34	1.9
	8	0.26	29	0.4
	9	0.17	40	0.1
Gold / DDDH	7	0.26	1	50
CPE[b]-DDDH	7	0.31	60	1.1

[a] SG – spectrographic graphite; [b] CPE – carbon paste electrode

The electrocatalytic activity of the DDD derivatives has been checked by means of cyclic voltammetry and estimated quantitatively by means of the rotating disc electrode technique. Importantly, it was shown that enzymatically active NAD^+ is indeed the main product of the oxidation, though the efficiency being quite low (ca. 50%) for DDDH immobilized on gold [80]. Disregarding the effect of the immobilization technique, the DDD derivatives show an electrocatalytic activity (see the values for k_{obs} and the values for the formal standard potential in Table 1) comparable to most of the redox dyes (10^2-10^4 M^{-1} s^{-1}), though being still lower than that reported for the fastest $2e^-/(1\text{-}2)H^+$ acceptors, e.g. thionine (1.2×10^5 $M^{-1}s^{-1}$, $E^{o\prime}$ = 0.24 V *vs.* Ag/AgCl) [82], p-benzoquinone (2.9×10^6 M^{-1} s^{-1}, $E^{o\prime}$ = 0.115 V *vs.* SCE) [83], and TCNQ (4.8×10^6 $M^{-1}s^{-1}$, $E^{o\prime}$ = 0.36 V *vs.* ECS) [84], 1,4-bis(3,4-dihydroxyphenyl)-2,3-dimethylbutane (3.6×10^5 $M^{-1}s^{-1}$, $E^{o\prime}$ = 0.14 V *vs.* Ag/AgCl) [85]. Apparently, the DDD derivatives are able to decrease significantly the overpotential of NADH oxidation. The ability to react with NADH at low overpotential can be understand in terms of decreasing the overpotential of the first electron transfer, which is probably accompanied by formation of a charge-transfer complex:

$$\text{DDDH} \leftrightarrows \text{DDD}^{\bullet} + e^- + H^+ \quad (13)$$

$$\text{DDD}^{\bullet} + \text{NADH} \leftrightarrows [\text{DDD}^{\bullet}\ \text{NADH}] \quad (14)$$

$$[\text{DDD}^{\bullet}\ \text{NADH}] \rightarrow [\text{DDDH NAD}^{\bullet}] \quad (15)$$

After the first electron/proton transfer (Eq. 14), the mediator molecule adopts a relatively stable electronic configuration (reduced form), thus losing the "connection" with the coenzyme. Indeed, it is not obvious that the [DDDH $NAD^{\bullet}$] complex is stable and would not dissociate prior to the re-oxidation of DDDH. The relatively low rate of the adsorbed DDDH oxidation brings about an increase of the $NAD^{\bullet}$ radical "life expectancy", favoring the $NAD^{\bullet}$ side reactions (e.g. $NAD^{\bullet}$ dimerization (K_d = 8×10^7 M^{-1} s^{-1} for dissolved radical [86]) or simply $NAD^{\bullet}$ diffusion into solution, ultimately decreasing the catalytic efficiency. Significantly, steady-state current measurements indicate gradual decrease of the electrocatalytic activity of the Au/DDDH modified electrode. However, after the NADH containing solution is exchanged with the supporting electrolyte solution and the electrode is tested voltammetrically, it is clear that the electrode "fouling" is reversible [80].

As a final remark, the above discussion gives a good illustration to the complexity and importance of kinetic studies of the electrocatalytic oxidation of NADH. Indeed, a thorough understanding of the factors controlling the efficiency and rate of the electron transfer is a prerequisite for the development of an efficient amperometric biosensor.

3. PEROXIDASE-BASED AMPEROMETRIC BIOSENSORS

Peroxidases form a class of redox enzymes, which are able to catalyze efficiently the oxidation of various reductive substrates (AH_2) by peroxides:

$$ROOH + AH_2 \xrightarrow{\text{peroxidase}} A + ROH + H_2O \quad (16)$$

Peroxidases are specific toward reduction of hydrogen peroxide and some inferior mono-substituted (methyl, ethyl) peroxides, and quite nonspecific concerning the reductive substrate used. Mono- and polyphenols, amines, diamines, some amino acids, metallic complexes (e.g. ferrocenes, $[Fe(CN)_6]^{4-}$, $[Ru(NH_3)_5py]]^{2+}$, $[Os(bpy)_2py]^{2+}$), quinones, cytochrome-c or some dyes in their leuco forms are examples of such reductive agents [87-89]. Peroxidases are ubiquitous in animals, plants and microorganisms. Their physiological role is not fully understood. Plant peroxidases contain protoheme IX as the prosthetic group, which is actually responsible for their dark red (to brown) colour. The Horseradish peroxidase is representative for its class and certainly the most studied peroxidase. Most of the fundamental research of peroxidases was carried out with this enzyme. Lately, more and more studies are focused on other plant peroxidase. The cytochrome-c peroxidase from yeast is the first peroxidase which three-dimensional structure was established [90].

Horseradish peroxidase (HRP, EC 1.11.1.7, mol. mass 42-45 kDa) consists of a glucoprotein (308 amino acid fragments), two calcium ions and a heme group bound non-covalently to the proteinic chain [91]. Remarkably, in contrast to many heme proteins, binding between the apoprotein and the heme group in HRP is broken relatively easy and reversibly. Commercially available HRP is in fact a mixture of a number of iso-enzymes, HRP-C being the dominant component. HRP-C is a cationic protein with p*I* of ca. 8.9 [92]. In normal conditions, native HRP is quite stable in a wide pH range (from 5 to 10). Its remarkable chemical stability and the multiple transduction schemes, which may be used in the amperometric biosensors design, make HRP an ideal enzyme for fundamental and applied bioelectrochemical research.

The biocatalytic cycle of the peroxidase can be presented schematically as follows:

$$HRP\ (Fe^{3+}) + H_2O_2 \xrightarrow{k_1} \text{Comp. I} + H_2O \quad (17)$$

$$\text{Comp. I} + AH_2 \xrightarrow{k_2} \text{Comp. II} + HA^{\bullet} \quad (18)$$

$$\text{Comp. II} + AH_2 \xrightarrow{k_3} HRP\ (Fe^{3+}) + HA^{\bullet} + H_2O \quad (19)$$

In simple terms, reaction (17) is a fast oxygen atom transfer to the prosthetic group (two-electron process), resulted in formation of the oxyferryl π-cation radical heme ($[Fe^{4+}{=}O]^{\bullet+}$) [91]. The oxidized form of the enzyme (also called compound I, Comp. I) is reduced stepwise (Eqs. 18 and 19) to the ferric enzyme, HRP(Fe^{3+}), through an intermediate species called compound II, Comp. II. At the same time, the hydrogen-donating reductive substrate is oxidized to the corresponding free radicals [93]. The metallic complexes and iodides are oxidized through one- and two-electron transfers, respectively, the proton being transferred from other species [94]. To summarize, the catalytic cycle of the peroxidase include a two-electron oxidation of the enzyme prosthetic group, followed by two consecutive $1e^-/1H^+$ transfers. Excessive concentration of the peroxide brings about formation of an enzymatically inactive compound, referred to as compound III, Comp. III [95].

The mechanism of peroxidase action is commonly viewed in terms of the so-called "ping-pong" mechanism, also referred to as the "parallel lines" kinetics [96]. If the initial reaction rate v is defined as $d[AH_2] / dt$, then the following expression can be derived from the reaction scheme (17-19) [91]:

$$\frac{2[\mathrm{HRP}]_o}{v} = \frac{(k_2 + k_3)}{k_2 k_3}\frac{1}{[\mathrm{AH_2}]} + \frac{1}{k_1[\mathrm{H_2O_2}]} \tag{20}$$

For most cases k_2 exceeds k_3 ca. 10 times. Therefore, equation (20) can be simplified as follows:

$$\frac{2[\mathrm{HRP}]}{v} = \frac{1}{k_3[\mathrm{AH_2}]} + \frac{1}{k_1[\mathrm{H_2O_2}]} \tag{21}$$

If the concentration of hydrogen peroxide is kept constant, then the rate constants k_1 and k_3 can be calculated from the intercept and slope, respectively, of the ($2[HRP]_o / v$) vs. ($1 / [AH_2]$) plot, which would give a set of parallel lines corresponding to different concentrations of the peroxide. At the same time the intercept of the primary function (*IPF*) has a zero intercept.

$$IPF = \frac{1}{k_1[\mathrm{H_2O_2}]} \tag{22}$$

This is a particularity of the mechanism of the peroxidase action: no K_M (Michaelis constant) can in principle be defined, since the catalytic cycle is irreversible. Finally, the rate constants k_1 and k_3 can be estimated from the steady-state measurements, but k_2 cannot.

We shall turn now to considering the electrical communication between the peroxidase active center and the electrode surface, particularly in relation with the amperometric detection of hydrogen peroxide. The peroxidase-based modified electrodes can be used for detection of compounds, which play role of a substrate, activator or inhibitor of the biocatalytic process. The compound I can be reduced to the enzymatically active ferric HRP via (i) mediated electron transfer or (ii) via heterogeneous electron transfer from the electrode surface. In both cases the resulted current density is a function of the hydrogen peroxide concentration.

The simplest peroxidase-based electrode for detection of hydrogen peroxide is obtained by placing (e.g. by adsorption) a monolayer of enzyme molecules on the surface of an electrode. When the modified electrode is introduced into an electrolyte solution and a sufficiently negative potential is applied (usually more negative than 0.6 V vs. SCE), the observed electrocatalytic current proves to be proportional to the concentration of hydrogen peroxide, in a certain concentration range. This phenomenon was observed for such electrode materials as graphite [97-101], carbon black [102], gold [102,103] and platinum [104,105] electrodes. It was concluded that the observed current is a result of a direct electron transfer to compounds I and II, which occurs at the electrode surface. In accordance with this conclusion, the electrocatalytic reduction of hydrogen peroxide at peroxidase-modified electrodes is observed at potentials close to 0.63-0.69 V vs. SCE (pH 7) range, in which the formal potentials of compounds I and II are situated [106,107]. More recently, it was shown that bioelectrocatalytic reduction of hydrogen peroxide can be achieved at carbon electrodes modified with cytochrome-c peroxidase as well as with a number of plant peroxidases [36,89].

The direct electrochemical reduction of the oxidized enzyme is sluggish at conventional electrode materials, the rate of the process increasing with the decrease of the applied potential. For HRP adsorbed on the spectral graphite, the heterogeneous rate constant was estimated at ca. 0.66 s^{-1} (at 0 V vs. SCE) [101]. The heterogeneous electron transfer seems to be much slower than the homogenous reduction by the peroxidase reductive substrate. Accordingly, the turnover for the homogeneous reduction of the compound II by the most active reductive substrates (e.g. *o*-dianisidine) ranges from 1000 to 2000 s^{-1} [101]. At the same time, low turnovers for the direct electrochemical reduction of the compound II may also be a result of a kinetic heterogeneity of the enzyme population. Thus, Ruzgas et al. [101] estimated that only ca. 40% of the adsorbed HRP molecules are actually involved in the electron exchange with the electrode surface (graphite), probably due to deactivation phenomena or inadequate orientation of the enzyme.

The use of a redox mediator, which plays the role of a reductive substrate, is an alternative to the peroxidase recycling via a direct electron transfer. A successful mediator would assure a fast and efficient reduction of the oxidized enzyme. Phenols, aromatic amines, ascorbic acid, iodides of the alkaline metals are able to reduce fast compounds I and II. Curiously enough, one-electron mediators, particularly metallic complexes, such as ferrocenes, hexocyanoferrate (II), some pyridine and bi-pyridine complexes of osmium, can assure a reasonably fast and efficient recycling of the peroxidase redox center. The second order rates constants of the peroxidase reduction by one-electron mediators (around 1×10^5 $M^{-1}s^{-1}$ for ferrocenes [108-110], 1.8×10^5 $M^{-1}s^{-1}$ for hexacyanoferrate (II) [111], 1.2×10^4 $M^{-1}s^{-1}$ for pentaamine-pyridine-ruthenium(II) complex [112], 7.8×10^2 $M^{-1}s^{-1}$ for bipyridine-osmium(II) complex [113]) are still lower than those for the electron-proton donors, such as hydroquinones (around 1×10^7 $M^{-1}s^{-1}$) or *p*-cresol (1×10^6 $M^{-1}s^{-1}$) [114]. Therefore, when one-electron mediators are employed, higher concentration of the metallic complexes is required for the transducing interface to reach the diffusion-limited regime of response. The rate of the compounds I and II reduction by metallic complexes is significantly lower than the rate of the reaction between the peroxidase and the hydrogen peroxide (between ca. 1×10^7 and 3×10^7 $M^{-1}s^{-1}$) [115].

In regard to the electrode configuration, the peroxidase-based modified electrodes for detection of hydrogen peroxide can be classified in three main groups: (i) the electrodes, which surface is modified with a monolayer of the peroxidase molecules immobilized via physical adsorption or via formation of covalent bonds with the surface; (ii) the electrodes modified with a polymer film, either redox or a conductive polymer, which contains the entrapped peroxidase molecules; (iii) the electrodes based on the composite materials, the peroxidase molecules being a part of the electrode material [89]. Stable electrode configuration can be obtained by either physical adsorption [116-118] or, more properly, by covalent binding via carbodiimide coupling or covering the electrode with a semi-permeable membrane [62,119-122]. The modified electrodes based on using the carbon paste represent an interesting alternative to the electrodes mentioned above, particularly in regard to the possibility of tuning the electrode characteristics through a proper choice of the stabilizer and/or mediator, and an easy renewability of the electrode surface [62,123-125]. The procedure of the peroxidase entrapment plays an important, if not decisive, role in determining the electrode performance, particularly when a direct electron transfer is aimed at. Thus, in order to avoid the enzyme isolation by the oil, employed in the preparation of the carbon paste, the enzyme must be first brought in contact with the carbon powder and then the oil [126]. Systematic studies of the effect of different additives and graphite powders showed that (i) the stability of the resulted modified electrodes can be improved and (ii) the sensitivity of the carbon paste electrodes based on the direct electron transfer can be increased up to 10-17 times by a proper choice of the additive and the source of the graphite powder [125].

An extensively studied and perhaps the most promising class of the peroxidase-modified electrodes is based on using conducting polymers (e.g. polypyrrole) or highly water soluble redox polymers (e.g. polyvinyl imidazole with $[Os(bpy)_2Cl]^{+/2+}$ redox center attached). In such electrodes, the polymer backbone (i) plays the role of immobilization matrix and (ii) electrically connects the peroxidase redox center with the electrode surface.

Electrocatalytic response of the electrodes containing the peroxidase molecules entrapped in the three-dimensional matrix of a conductive polymer (e.g. electropolymerized [127] or potentiostatically-deposited [118] polypyrrole, electropolymerized o-phenylendiamine [128], water-soluble poly(3-(3′-thienyl)propanesulfonic acid) [129,130] was interpreted in terms of electron transfer from the electrode surface to the redox center of the enzyme via the electron conducting polymer matrix. Accordingly, the polymer matrix may be viewed as an extension of the electrode surface. The relatively low sensitivity of the electrodes based on the conducting polymers is related to a low diffusivity of hydrogen peroxide in the polymer film [127,131]. The structural integrity of the polymer and the conduction mechanism may have certain effect on the bio-electrode response as well as the mechanism of the electrical communication. For example, there is enough evidence pointing to the importance of a mediated electron transfer for HRP entrapped in a polypyrrole film, due to the presence of the oxidized pyrrole oligomers [127,129]. Furthermore, higher current densities are observed at lower (acidic) pH values, probably related to the restricted mobility of the protons within the film.

Electrodes modified by covering with redox or electron conducting polymers, incorporating physically or chemically attached enzyme molecules, represent an important and extensively studied group of peroxidase-based electrodes [89]. Thus, electrodes based on HRP entrapped into $[Os^{III/II}(bpy)_2Cl]$-containing polyvinyl pyridine or polyvinyl imidazole films [113,132,133] are among the most successful peroxidase-based modified electrodes, if such criteria as sensitivity, lower detection limit and stability are considered (for comparison see Table 2 in reference 89). Note, that the peroxidase-modified electrodes based on using redox polymers are among (relatively few) commercially viable amperometric bio-analytic devises (Bioanalytical Systems, West Lafayette, IN, cat. Nr. MF-2095 and MF-2098).

The above-mentioned polymer materials are commonly referred to as redox "wires" or redox hydrogels. A redox "wire" is a redox macromolecule designed to: (i) physically attach the enzyme to the electrode; (ii) strongly complex the enzyme protein; (iii) electrically connect the enzyme's redox center to the electrode, all these without serious deactivation of the enzyme [134] When an oxidoreductase is integrated in such a three-dimensional polymer matrix, electrons are transported *via* the redox polymer network between the enzyme's redox center and electrode. Extended experimental data on the electrical communication between the enzyme and redox polymers is now available [132,134-138], although the subject is far to be exhausted. Thus, the kinetic aspects of the enzymes "wiring" by redox polymers are particularly interesting [134].

In early 1990s, it was demonstrated that uniform thin polymeric films can be assembled on a variety of substrates by alternating electrostatically controlled adsorption of polycations and polyanions on charged surfaces [10]. In a short time, layer-by-layer assembly technique proved to be an efficient strategy for the fabrication of multicomponent nanostructured films on various solid supports [139]. Much of the recent work on these polyelectrolyte multilayers has been focused on generalizing and expanding the technique to various combinations of charged components, including synthetic polyelectrolytes [139], biopolymers, such as proteins [140,141] and enzymes [138,142-150]. Most importantly, building of multilayer redox polyelectrolyte-oxidoreductase architectures, based on electrostatic interactions, allows fine control of the film thickness and the amount of the biomolecule at the electrode surface and, from these considerations, the technique could present a promising alternative to electrode coating with thick redox hydrogel films loaded with enzyme molecules.

In a recent paper [151], the potentiality of multilayer self-deposition technique based on electrostatic interactions for the enzyme bioelectrodes design was analyzed. The authors examine reagentless biosensors build up by alternate layer-by-layer self-deposition of positively charged poly[(vinylpyridine)$Os^{III/II}(bpy)_2Cl$] redox polymer (RP) layer(s) and some oxidoreductases, including chemically modified HRP, on a negatively charged alkanethiolate-modified gold electrode surface. The sensitivity of 58.1 mA M^{-1} cm^{-2} observed for one RP-HRP_{mod} pair of layers was much lower than that of 1 A M^{-1} cm^{-2} – theoretically maximal sensitivity of a peroxidase electrode [89] and actually the highest value of the sensitivity ever reported for the hydrogen peroxide detection at HRP-based bioelectrodes [132]. A straightforward way to increase the sensitivity is further alternate deposition of new biocatalytic and transducing layers, resulting in a multilayer assembly. This could bring about an increase of the importance of both electron "diffusion" and the substrate diffusion limitations. Consequently, only a thorough kinetic analysis of the resulted multilayer RP-HRP assembly, as well as its structural elements, could provide further insight into the interface and could be of help in identification of certain criteria for optimization of the (bio)electrode response.

There are few studies [138,149,150] addressing the kinetics of the peroxidase "wiring" (immobilized mediator!) and the identification of the kinetic parameters or any other factors controlling the efficiency of the electrical communication between the peroxidase active center and the electrode surface. The approach is essentially based on an expression describing the dependence of the catalytic current on the substrate concentration, when the substrate diffusion is facile [152,153]. Combining the kinetic parameters for the soluble enzyme (the apparent Michaelis-Menten constant and the turnover number) with those furnished by the non-linear fit of the above expression to the experimental data, one can estimate the enzyme surface concentration and the pseudo-first-order rate constant corresponding to the reaction between the enzyme and the redox polymer in the investigated architecture [138,154]. However, the variation of the enzyme kinetic parameters due to its immobilization on the electrode surface or participation of only a fraction of the (adjacent) mediator redox centers in the electron exchange with the enzyme should to be taken into consideration as well. More recently, Calvente et al. [154,155] reported a detailed theoretical analysis of the kinetics of the immobilized HRP "wiring" in a redox polymer film, including the substrate diffusion through the polymer film and the charge transport (electron "hopping") through the film. Starting from the exact analytical solutions for a two-dimensional redox mediator-enzyme configuration, the authors showed that under certain experimental conditions (most importantly, uniform redox conversion throughout the film) the same equations can be used for analyzing the three-dimensional films. Then, the authors concluded that (i) the electrical communication within the redox polyelectrolyte-peroxidase film is limited by the conversion of the HRP-H_2O_2 complex into compound I (the authors assume here the Michaelis-Menten formalism) and (ii) the mediating steps and the substrate diffusion through the film are facile. Finally, the authors suggest that the main issue in the optimization of the systems investigated would be an increase of the fraction of the enzyme molecules involved into the bioelectrocatalytic cycle.

The Michaelis-Menten (M-M) formalism was commonly employed in early studies on peroxidase based electrodes. Accordingly, the bioelectrocatalytic response was often characterized by the apparent M-M constant K_M' and the apparent maximum current i_{max}', which provided suitable criteria for comparison of different peroxidase-modified electrodes [125,156]. Direct application of the M-M formalism to the case of peroxidase action is not always adequate. Accordingly, the oxidation of the ferric enzyme, HRP(Fe^{3+}), is a fast irreversible process [91]. The same applies to the reduction of the oxidized intermediates compound I and compound II [91]. Nevertheless, the M-M formalism was employed in detailed kinetic studies of mediated (by an osmium complex) electrochemistry of HRP [157] and the HRP "wiring" by redox polymers [155]. On the other hand, an M-M-like expression can be derived for some model peroxidase-redox polymer interfaces starting with the "ping-pong" mechanism of the peroxidase action in the manner similar to that given by Tatsuma and Watanabe [158] and later applied by Bogdanovskaya et al. [156] to the peroxidase-modified electrodes involving direct electron transfer. Later, Gorton et al. showed that combining the "ping-pong" kinetics with the Koutecky-Levich formalism (RDE technique) allows estimation of a number of kinetic parameters for the case of mediated or direct electrochemical reduction of the oxidized HRP [36,101].

For illustrating how the two approaches can provide valuable kinetic information, let us consider in some detail a kinetic analysis of a model interface, which is presented schematically in Figure 5. The Au/MPS/RP/HRP_m/RP modified electrode was formed by sequential adsorption of cationic poly[(vinylpyridine)$Os^{III/II}$(bpy)$_2$Cl] redox polymer (RP) partially quaternised with bromoethylamine [159] and chemically modified [151] HRP (HRP_m) on gold electrodes modified with a self-assembled monolayer of 3-mercapto-1-

propane sulfonic acid (MPS) [160]. The chemical modification of the redox polymer and the enzyme were shown to promote the (electrostatic) interactions between the two macromolecules and improve the electrical communication within the thick film [151,159] or self-deposited assembly [151]. Note that the outermost RP layer was deposited in order to minimize the enzyme orientation effect. Finally, since electrostatic forces are perceived as the forces that control the assembly growth [151], HRP_m adsorption ought to be limited to a nearly single molecular layer.

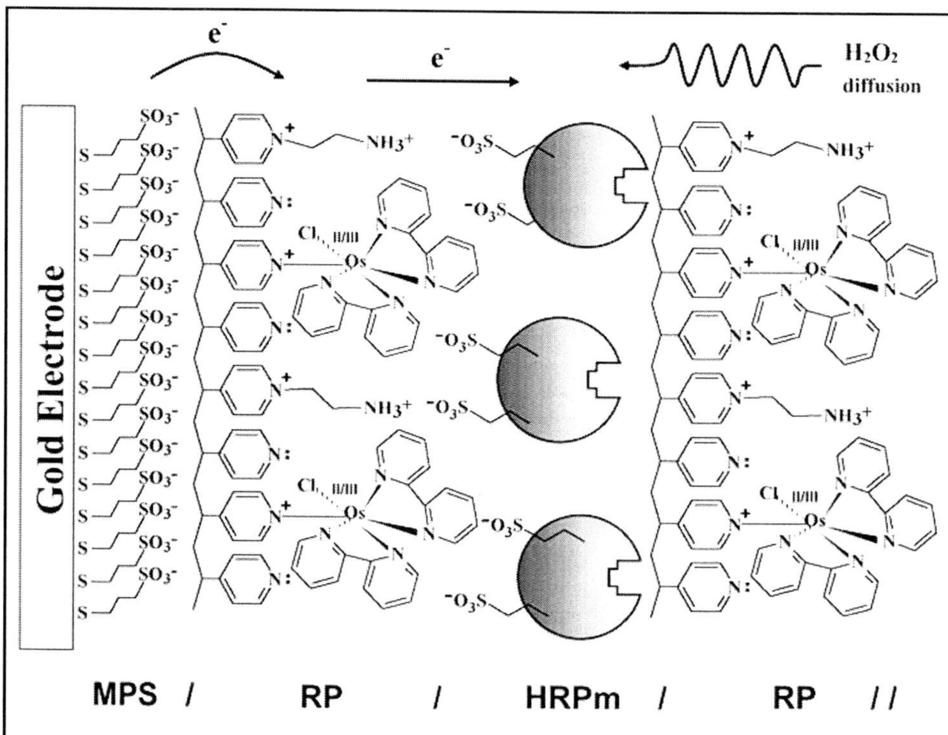

Figure 5. Schematic representation of the $Au/MP/RP/HRP_m/RP$ interface.

The following reaction sequence can be proposed for the case of mediated bioelectrocatalytic reduction of H_2O_2 at $Au/MPS/RP/HRP_m/RP$ electrode:

$$HRP_m\,(Fe^{3+}) + H_2O_2 \xrightarrow{k_1} \text{Comp. I} + H_2O \quad (23)$$

$$\text{Comp. I} + Os^{II}\,(+\,H^+) \xrightarrow{k_2} \text{Comp. II} + Os^{III} \quad (24)$$

$$\text{Comp. II} + Os^{II}\,(+\,H^+) \xrightarrow{k_3} HRP_m\,(Fe^{3+}) + Os^{III} + H_2O \quad (25)$$

$$(Os^{II})_1 + (Os^{III})_2 \xrightarrow{k_e} (Os^{III})_1 + (Os^{II})_2 \quad (26)$$

$$Os^{III} + e^- \xrightarrow{k_{S,M}} Os^{II} \quad (27)$$

Reactions (23), (24) and (25) corresponds to reactions (17), (18) and (19). Note the mediator inability to transfer protons, which means involvement of a proton-donating solution species. Reaction (26) describes transport of the oxidizing equivalents to the electrode surface by the self-exchange reaction (electron "hopping"). Fast electron transfer (reaction 27) from the electrode surface to the oxidized mediator closes the electrocatalytic cycle.

Assuming the stationary state for Comp.I and Comp.II concentrations, we can write:

$$d\Gamma_{CpI} / dt = k_1\Gamma_{HRP(Fe3+)} [H_2O_2] - k_2\Gamma_{CpI} [Os^{II}] = 0 \qquad (28)$$

$$d\Gamma_{CpII} / dt = k_2\Gamma_{CpI} [Os^{II}] - k_3\Gamma_{CpII} [Os^{II}] = 0 \qquad (29)$$

Additionally, the mass balance for all enzyme species requires:

$$\Gamma_{tot} = \Gamma_{HRP(Fe3+)} + \Gamma_{CpI} + \Gamma_{CpII} \qquad (30)$$

where Γ_{CpI}, Γ_{CpII} and $\Gamma_{HRP(Fe3+)}$ are the effective (involved in bioelectrocatalytic cycle) surface concentrations of the corresponding forms of the enzyme (mol cm^{-2}); $[H_2O_2]$ stands for the bulk concentration of hydrogen peroxide (mol l^{-1}); $[Os^{II}]$ is the effective volume concentration of the osmium complex in the reduced form (mol l^{-1}). Expressing Γ_{CpI} and Γ_{CpII} from relation (28) and (29), respectively, and then substituting the resulting relations in equation (30) gives:

$$\Gamma_{tot} = \Gamma_{HRP(Fe3+)} \{ 1 + (k_1[H_2O_2] / k_2[Os^{II}]) + (k_1[H_2O_2] / k_3[Os^{II}]) \} \qquad (31)$$

Next, the catalytic current at the electrode under investigation is determined by the rate of reactions (24) and (25), since the osmium complex is a sluggish reductive substrate (see also the discussion below):

$$i_{cat} = F (k_2\Gamma_{CpI} [Os^{II}] + k_3\Gamma_{CpII} [Os^{II}]) \qquad (32)$$

Combining expressions for Γ_{CpI} and Γ_{CpII} from relation (28) and (29), respectively, with equation (32), then multiplying the resulting expression by equation (31), and performing few simple mathematical transformations the following expression for the electrocatalytic current results:

$$i_{cat} = (2F k_1[H_2O_2]\Gamma_{tot}) / \{ 1 + [H_2O_2] (k_1 / [Os^{II}]) [(k_2 + k_3) / (k_2 k_3)] \} \qquad (33)$$

Denoting i_{max} and K_M as:

$$i_{max} = (2F k_2 k_3 [Os^{II}]\Gamma_{tot}) / (k_2 + k_3) \qquad (34)$$

$$K_M = [Os^{II}] / [k_1 (k_2 + k_3) / (k_2 k_3)] \qquad (35)$$

relation (33) takes an M-M-like form:

$$i_{cat} = (i_{max} [H_2O_2]) / (K_M + [H_2O_2]) \qquad (36)$$

At this stage it would be useful to discuss the significance of this result. First of all, eqs. (34) and (35) are essentially similar to those derived by Bogdanovskaya et al. [156] for the case of the peroxidase electrodes based on a direct electron transfer. This result is quite logical, since the interface considered contains one monolayer of the enzyme. Secondly, though the presence of $[Os^{II}]$ is formally correct, it should be acknowledged that this kinetic formalism would be of little use for elucidating the effects related to the electron transfer between the enzyme and the redox centers of the transducing macromolecule. Accordingly, the efficiency of the enzyme reduction by the redox polymer would be determined not simply by the effective (local) concentration of its redox canters, but would be also influenced by other factors, such as the orientation effects; nature and concentration of the supporting electrolyte; linear dimensions of the enzyme and the reductive redox center; and so on. Thirdly, the formalism presented would be inadequate for kinetic analysis of the multilayer assemblies, where three-dimensional distribution of the enzyme content, the diffusion of H_2O_2 through the assembly and eventually charge transport (reaction 26) through the assembly should be also considered.

In spite of its simplicity, the model applied presents a formally correct approach for estimating the enzyme activity. Accepting $k_2 >> k_3$, [161] expression (34) reduces to the following:

$$i_{max} = 2F\, k_3\, [Os^{II}]\Gamma_{tot} \qquad (37)$$

Now, by combining (35) with (37) one can easily derive the expression, which allows calculus of k_1:

$$i_{max} / K_M = 2Fk_1\Gamma_{tot} \qquad (38)$$

Non-linear regression analysis of the i_{cat} vs. $[H_2O_2]$ dependence, recorded under the kinetic control (RDE, 2400 rpm) gave a value of 3.7×10^{-6} A for i_{max} and 82×10^{-6} M for K_M. From equation (38), a value of 1.49×10^{5} $M^{-1}s^{-1}$ was obtained for the second order rate constant of the reaction between H_2O_2 and HRP (Γ_{HRP} was estimated [160] at 1.8 pmol cm^{-2}).

Turning now to the effect of the applied potential, we do not observe any significant effect of the applied potential (region between 0-0.3 V *vs.* Ag|AgCl|KCl_{sat}, pH 7.0) on the electrocatalytic current observed at Au/MPS/RP/HRP_m/RP [162], while in the case of the modified (carbon) electrodes peroxidase based on a direct electron transfer the effect is commonly noticed [101,125,156,158]. The effect of the applied potential observed for the peroxidase-modified carbon electrodes may be tentatively explained by the conformation changes in the adsorbed peroxidase, probably resulted in shorter electron transfer distances. Furthermore, Bogdanovskaya et al. [156] assigned the positive effect of more negative potentials to an increase of the rate constant k_3, as this corresponds to an electrochemical step. In contrast, in the case of the electrodes based on the peroxidase "wiring", the reduction of the oxidized enzyme is a chemical process, hence no direct effect of the applied potential is expected. Additionally, the lack of the effect of the applied potential on the investigated interface response would be in agreement with the conclusion that neither heterogeneous electron transfer between the Os redox center and the electrode surface nor the charge transport through the assembly (electron hopping) are rate-determining steps in the examined potential region [160]. Note, that similar conclusions were drawn from analysis of the thick redox polymer-peroxidase films [132,134]. Tthe catalytic reduction of the hydrogen peroxide by the peroxidase is a fast process (15×10^{6} M^{-1} s^{-1}) [115], therefore, the regeneration of the enzyme, i.e., electron transfer between the Os reductive substrate and the oxidized forms of the enzyme, should determine the overall rate of the process. Furthermore, the $[Os^{II}(bpy)_2Cl]$ complex reacts slowly with the oxidized native HRP in solution (7.8×10^{2} M^{-1} s^{-1}) [113], thus being a quite sluggish reductive substrate for HRP.

Turning now back to the kinetic analysis, equation (33) can be rewritten as follows:

$$\frac{1}{i_{cat}} = \frac{1}{2FA\Gamma_{HRP}}\left(\frac{1}{k_1[\mathrm{H_2O_2}]} + \frac{k_2 + k_3}{k_2 k_3}\frac{1}{[\mathrm{Os^{II}}]}\right) \tag{39}$$

Note that this expression is identical to that derived for mediated bioelectrocatalytic reduction of H_2O_2 at (sub)monolayer HRP-modified electrode in presence of a freely diffusing electron donor substrate [36]. As mentioned earlier in this chapter, this approach is quite simplistic and would not allow estimation of k_2 or k_3 . In spite of this, the model allows estimation of the turnover number of the oxidized enzyme reduction by $[Os^{II}(bpy)_2Cl]$ redox centers. Thus, equation (39) can be rewritten as follows:

$$\frac{1}{i_{cat}} = \frac{1}{2FA\Gamma_{HRP}}\left(\frac{1}{k_1[\mathrm{H_2O_2}]} + \frac{1}{k_{S,ET}}\right) \tag{40}$$

The (apparent) turnover number $k_{S,ET}$ (s^{-1}) could be informative from two main reasons: (i) it is a parameter, which characterize the efficiency of the electrical communication between the enzyme and redox polymer, and, consequently, could be used as a criterion for optimization of the bioelectrode response; (ii) it allows direct comparison with other peroxidase-based bioelectrodes, particularly those based on direct electron transfer (ET) between the enzyme active center and the electrode surface.

In the the Koutecky-Levich formalism (RDE technique), the observed H_2O_2 reduction current (I) measured at the Au/MPS/RP/HRP_{mod}/RP electrode is a combination of the mass-transport-limited current I_L (the second right-side term) and the reaction-rate-limited current I_K [163]:

$$\frac{1}{I} = \frac{1}{I_K} + \frac{1}{0.62nFAD^{2/3}\omega^{1/2}\nu^{-1/6}[\mathrm{H_2O_2}]} \tag{41}$$

Equation (40) involves the reaction-rate-limited current I_K, which, in fact, is equal to the observed (biocatalytic) current (i_{cat}), provided the substrate diffusion is facile (high rotation speed of the RDE). By plotting I^{-1} *versus* $\omega^{-1/2}$ the reaction-rate-limited current can be separated from the mass-transport-limited one. From the slope and intercept of the dependence of i_{cat} on $[H_2O_2]$ in doubly reciprocal plot, k_1 and $k_{S,ET}$, respectively, can be calculated.

Using the value of 8 pmol cm^{-2} as an estimation [160] of Γ_{HRP}, the rate constant k_1 of the reaction between H_2O_2 and HRP_{mod} was found to be 2.74×10^5 M^{-1} s^{-1} and the apparent turnover number $k_{S,ET}$ was equal to 2.14 s^{-1}. The rate constant k_1 for HRP_{mod} in RP/HRP_{mod}/RP architecture is about 50 times lower than that for the wild type HRP in solution (15×10^6 M^{-1} s^{-1}) [115]. This considerable difference could be assigned to the effect of the enzyme immobilization and probably to the enzyme partial deactivation upon its chemical modification.

As already mentioned, $k_{S,ET}$ allows a direct comparison between the investigated transducing scheme and the schemes based on direct electron transfer between the enzyme active center and electrode. Thus, the value calculated for HRP_{mod} (2.14 s^{-1}) is close, but somehow higher than, to the turnover number of the heterogeneous electron transfer for wild type HRP adsorbed on spectroscopic graphite (0.66 s^{-1}; pH 7 [101]) and is considerably higher than that at a graphite coating (2.0×10^{-4} s^{-1}, pH 6.4; [164]). In this context, the sensitivity of the Au/MPS/RP/HRP_{mod}/ interface (58.1 mA M^{-1} cm^{-2}) [151] is comparable to that observed for the pyrolytic graphite electrodes modified with physically adsorbed native HRP (60 mA M^{-1} cm^{-2}) [118]. In both cases, the enzyme is most probably presented in a nearly single molecular layer. These facts suggest that the efficiency of the peroxidase regeneration by the transducing redox polymer is most probably superior but still comparable to that provided by direct electron transfer. At the same time, to be recalled and emphasized the fundamental advantage of using redox polymer networks. These conducting immobilization matrices provide electrical communication for enzyme molecules situated far from the surface, effectively increasing the "virtual" surface of the electrode [16] and ultimately resulting in superior analytic characteristics of the resulted biosensors [89,132]. Furthermore, the "wiring" efficiency may be improved by manipulating both transducing macromolecule and the enzyme.

As a final remark on the peroxidase "wiring" with redox polymers, a fundamental understanding of the processes responsible for signal transduction in redox polymer – peroxidase assemblies (films) requires a detailed mechanistic and kinetic analysis, based on a rigorous mathematical description of the system. At the same time, a more practice-oriented approach, like the one presented and discussed in previous paragraphs, may provide with valuable criteria for evaluating the "wiring" efficiency and, ultimately, for the biosensor optimization. Accordingly, once a carefully characterized redox polymer (narrow distribution of the molecular weight and an uniform mediator loading) and the enzyme are available, one can systematically study the effect of the electrolyte composition (ionic force, pH etc.) on a model interface, based on the estimation of the enzyme activity and the turnover of its reduction. Then, the effect of charge of the transducing polymer and the enzyme, the enzyme loading, mediator loading and the nature of the mediator group may be studied quite straightforwardly, since the factors mentioned may in principle affect the peroxidase apparent catalytic activity and the apparent turnover of its regeneration.

REFERENCES

[1] Turner, A. P. F. In *Biosensors - Fundamentals and Applications*; Turner, A. P. F., Karube, I., Wilson, G. S., Eds.; Oxford University Press: Oxford, 1987.

[2] Buerk, D. G. *Biosensors. Theory and Applications*; Technomic Publishing Co., Inc.: Lancaster, 1993.

[3] Lowe, C. R. *Biosensors* 1985, *1*, 3.

[4] Goepel, W.; Heidruschka, P. *Biosens. Bioelectron.* 1995, *10*, 853.

[5] Schmidt, H.-L.; Schumann, W.; Scheller, F. W.; Schubert, F. In *Sensors. A Comrehensive Survey*; Goepel, W., Hesse, J., Zemel, J. N., Eds.; VCH: Weinheim, 1992; Vol. 3.

[6] Handbook of Chemical and Biological Sensors; Taylor, R. F., Schultz, J. S., Eds.; Institute of Physics Publishing: Bristol, Uk, 1996.

[7] Goepel, W. *Biosens. Bioelectron.* 1995, *10*, 35.

[8] Bartlett, P. N. *Medical & biological engineering & computing* 1990, *28*, B10.
[9] Molecular Design of Electrode Surfaces; Murray, R. W., Ed.; Wiley: New-York, 1992.
[10] Decher, G.; Schmitt, J. *Prog. Colloid. Polym. Sci.* 1992, *89*, 160.
[11] Schuhmann, W. *ACS Symposium Series* 1994, *556*, 110.
[12] Allara, D. L. *Biosens. Bioelectron.* 1995, *10*, 771.
[13] Chaudhurry, M. K. *Biosens. Bioelectron.* 1995, *10*, 785.
[14] Ratner. *Biosens. Bioelectron.* 1995, *10*, 797.
[15] Willner, I.; Katz, E.; Willner, B. *Electroanalysis* 1997, *9*, 965.
[16] Habermuller, K.; Mosbach, M.; Schuhmann, W. *Fresenius' J. Anal. Chem.* 2000, *366*, 560.
[17] Palmisano, F.; Zambonin, P. G.; Centonze, D. *Fresenius' J. Anal. Chem.* 2000, *366*, 586.
[18] Willner, I.; Katz, E. *Angewandte Chemie, International Edition* 2000, *39*, 1181.
[19] Rosenwald, S. E.; Kuhr, W. G. *Electroanalytical Methods for Biological Materials* 2002, 399.
[20] Vidal, J.-C.; Garcia-Ruiz, E.; Castillo, J.-R. *Microchim. Acta* 2003, *143*, 93.
[21] Goepel, W. *Sens. Actuators A* 1996, *56*, 83.
[22] Griffiths, D.; Hall, G. *Trends Biol. Technol.* 1993, *11*, 122.
[23] Berg, J. M.; Tymoczko, J. L.; Stryer, L. *Biochemistry*, 5th ed.; W.H. Freeman & Co.: New York, 2002.
[24] Jencks, W. P. *Catalysts in Chemistry and Enzymology*; McGraw-Hill Company: New York, 1969.
[25] Braunstein, A. E.; Journ, D. I. *Mendeleev All-Union Chemical Society* 1963, *8*, 8.
[26] Marcus, R. A.; Sutin, N. *Biochim. Biophys. Acta* 1985, *811*, 265.
[27] Jeuken, L. J. C. *Biochim. Biophys. Acta* 2003, *1604*, 67.
[28] Lobo, M. J.; Miranda, A. J.; Tunon, P. *Electroanalysis* 1997, *9*, 191.
[29] Schumann, W. *Rev. Mol. Biotech.* 2002, *82*, 425.
[30] Heller, A.; Degani, Y. *J. Chem. Soc.* 1988, *110*, 2615.
[31] Whitesides, G. M. *Science* 1991, *254*, 1312.
[32] Loetzbeyer, T.; Schuhmann, W.; Schmidt, H.-L. *Sens. Actuators B* 1996, *33*, 50.
[33] Loetzbeyer, T.; Schuhmann, W.; Schmidt, H.-L. *J. Electroanal. Chem.* 1994, *377*, 377.
[34] Willner, I.; Helegshabtai, V.; Blonder, R.; Katz, E.; Tao, G. L. *J. Am. Chem. Soc.* 1996, *118*, 10321.
[35] Ghindilis, A. L.; Atanasov, P.; Wilkins, E. *Electroanalysis* 1997, *9*, 661.
[36] Gorton, L.; Linfgren, A.; Larsson, T.; Munteanu, F. D.; Ruzsgas, T.; Gazaryan, I. *Anal. Chim. Acta* 1999, *400*, 91.
[37] Armstrong, F. A.; Wilson, G. A. *Electrochim. Acta* 2000, 2623.
[38] Fertsh, A. *Enzyme structure and Mechanism*; W.H. Freeman&Co., 1977.
[39] Katakis, I.; Dominguez, E. *Mikrochim. Acta* 1997, *126*, 11.
[40] Lehninger, A. L. *Biochemistry*, 2nd ed.; Worth: New-York, 1975.
[41] Burnett, J. N.; Underwood, A. L. *Biochemistry* 1975, *5*, 2060.
[42] Jaegfeldt, H.; Torstensson, A.; Johansson, G. *Anal. Chim. Acta* 1978, *97*, 221.
[43] Blaedel, W. J.; Haas, R. G. *Anal. Chem.* 1970, *42*, 918.
[44] Moiroux, J.; Elving, P. J. *J. Electroanal. Chem.* 1979, *102*, 93.
[45] Moiroux, J.; Elving, P. J. *J. Am. Chem. Soc.* 1980, *102*, 6533.
[46] Blankespoor, R. L.; Miller, L. L. *J. Electroanal. Chem.* 1984, *171*, 231.

[47] Jaegfeldt, H. *J. Electroanal. Chem.* 1980, *110*, 295.

[48] Samec, Z.; Elving, P. J. *J. Electroanal. Chem.* 1983, *144*, 217.

[49] Hapiot, P.; Moiroux, J.; Saveant, J.-M. *J. Am. Chem. Soc.* 1990, *112*, 1337.

[50] Fukusumi, S.; Koumitsu, S.; Hironaka, K.; Tanaka, T. *J. Am. Chem. Soc.* 1987, *109*, 305.

[51] Carlson, B. W.; Miller, L. L. *J. Am. Chem. Soc.* 1983, *105*, 7453.

[52] Carlson, B. W.; Miller, L. L.; Neta, P.; Grodkowski, J. *J. Am. Chem. Soc.* 1984, *106*, 7233.

[53] Matsue, T.; Suda, M.; Uchida, I.; Kato, T.; Akiba, U.; Osa, T. *J. Electroanal. Chem.* 1987, *234*, 163.

[54] Grodkowski, J.; Neta, P.; Carlson, B. W.; Miller, L. L. *J. Phys. Chem.* 1983, *87*, 3135.

[55] Martens, F. M.; Verhoeven, J. W. *Recl. Trav. Chim. Pays-Bas* 1981, *100*, 228.

[56] Anne, A.; Hapiot, P.; Moiroux, J.; Neta, P.; Saveant, J.-M. *J. Phys. Chem.* 1991, *95*, 2370.

[57] Kuhr, W. G.; Barrett, V. L.; Gagnon, M. R.; Hopper, P.; Pantano, P. *Anal. Chem.* 1993, *65*, 617.

[58] Bresnahan, W. T.; Moiroux, J.; Samec, Z.; Elving, P. J. *Bioelectrochem. Bioenerg.* 1980, *7*, 125.

[59] Blaedel, W. J.; Jenkins, R. A. *Anal. Chem.* 1975, *47*, 1337.

[60] Gorton, L. *J. Chem. Soc., Faraday Trans.* 1986, *82*, 1245.

[61] Bartlett, P. N.; Tebbutt, P.; Whitaker, R. G. *Prog. Reaction Kinetics* 1991, *16*, 55.

[62] Gorton, L.; Csoeregi, E.; Dominguez, E.; Emneus, J.; Jinsson-Pettersson, G.; Marko-Varga, G.; Persson, B. *Anal. Chim. Acta* 1991, *250*, 203.

[63] Kitani, A.; Miller, L. L. *J. Am. Chem. Soc.* 1981, *103*, 3595.

[64] Kitani, A.; So, Y.-H.; Miller, L. L. *J. Am. Chem. Soc.* 1981, *103*, 7636.

[65] Carlson, B. W.; Miller, L. L. *J. Am. Chem. Soc.* 1985, *107*, 479.

[66] Fukuzumi, S.; Nishizawa, N.; Tanaka, T. *J. Org. Chem.* 1984, *49*, 3571.

[67] Gorton, L.; Persson, B.; Hale, P. D. In *Biosensors and Chemical Sensors. ACS Symposium Series.*; Edelman, P. G., Wang, J., Eds., 1992; Vol. 487; pp 56.

[68] Gorton, L.; Dominguez, E. *Rev. Molec. Biotech.* 2002, *82*, 371.

[69] Mano, N.; Kuhn, A. *J. Electroanal. Chem.* 1999, *477*, 79.

[70] Casero, O.; Darder, M.; Takada, K.; Abruna, H. D.; Pariente, F.; Lorenzo, E. *Langmuir* 1999, *15*, 127.

[71] Laviron, E. In *Electroanalytical Chemistry*; Bard, A. J., Ed.; Marcel Dekker: New York, 1982; Vol. 12; pp 53.

[72] Huck, H. *Phys. Chem. Chem. Phys.* 1999, *1*, 855.

[73] Andrieux, C. P.; Saveant, J. M. *J. Electroanal. Chem.* 1978, *93*, 163.

[74] Murray, R. W. In *Electroanalytical Chemistry*; Bard, A. J., Ed.; Marcel Dekker: New York, 1984; Vol. 13; pp 191.

[75] Popescu, I. C.; Dominguez, E.; Narvaez, A.; Pavlov, V.; Katakis, I. *J. Electroanal. Chem.* 1999, *464*, 208.

[76] Dicu, D.; Munteanu, F. D.; Popescu, I. C.; Gorton, L. *Anal. Lett.* 2003, *36*, 1755.

[77] Silberg, I. A.; Cristea, C. *Heterocycl. Comm.* 1996, *2*, 117.

[78] Silberg, I. A.; Silaghi-Dumitrescu, I.; Cristea, C.; Tordo, P.; Gigmes, D. *Heterocycl. Comm.* 1999, *5*, 147.

[79] Dicu, D.; Muresan, L.; Popescu, I. C.; Cristea, C.; Silberg, I. A.; Brouant, P. *Electrochim. Acta* 2000, *45*, 3951.
[80] Rosca, V.; Muresan, L.; Popescu, I. C.; Cristea, C.; Silberg, I. A. *Electrochem. Comm.* 2001, *3*, 439.
[81] Munteanu, F. D.; Dicu, D.; Popescu, I. C.; Gorton, L. *Electroanalysis* 2003, *15*, 383.
[82] Ohtani, M.; Kuwabata, S.; Yoneyama, H. *J. Electroanal. Chem.* 1997, *422*, 45.
[83] Murthy, A. S. N.; Sharma, J. *Talanta* 1998, *45*, 951.
[84] Murthy, A. S. N.; Anita, A.; Gupta, R. L. *Anal. Chim. Acta.* 1994, *289*, 43.
[85] Ciszewski, A.; Milczarek, G. *Anal. Chem.* 2000, *72*, 3203.
[86] Jensen, M. A.; Elving, P. J. *Biochim. Biophys. Acta* 1984, *764*, 310.
[87] Chance, B. *Arch. Bioch. Biophys.* 1949, *21*, 416.
[88] Saunders, B. C.; Holmes-Siedle, A. G.; Stark, B. P. *Peroxidase*; Butterworhts: London, 1964.
[89] Ruzsgas, T.; Csoeregi, E.; Emneus, J.; Gorton, L.; Marko-Varga, G. *Anal. Chim. Acta* 1996, *330*, 123.
[90] Poulos, T. L.; Freer, S. T.; Alden, R. A.; Edwards, S. L.; Skojlund, U.; Takio, K.; Eriksson, B.; Xuong, N.-H.; Yonetani, T.; Krant, J. *J. Biol. Chem.* 1980, *255*, 575.
[91] Dunford, H. B. Horseradish Peroxidase: Structure and Kinetic properties. In *Peroxidases in Chemistry and Biology*; Everse, J., Everse, K. E., Eds.; CRC Press: Boca Raton, 1992; Vol. 2; pp 2.
[92] Welinger, K. *Eur. J. Biochem.* 1997, *96*, 483.
[93] Dunford, H. B.; Adeniran, A. J. *Arch. Biochem. Biophys.* 1986, *251*, 536.
[94] Ugarova, N. N.; Lebedeva, O. V. *Biokhymyia* 1978, *43*, 1731.
[95] Adeiran, S. A.; Lambeir, A.-M. *Eur. J. Biochem.* 1989, *186*, 571.
[96] Purich, D. L. Criteria for evaluating the catalytic competence of enzyme-substrate covalent compounds. In *Contemporary Enzyme Kinetics*; Purich, D. L., Ed.; Academic Press: New York, 1983.
[97] Razumas, V.; Jasaitis, J.; Kulys, J. *Bioelectrochem. Bioenerg.* 1984, *12*, 297.
[98] Jonsson, G.; Gorton, L. *Electroanalysis* 1989, *1*, 465.
[99] Bogdanovskaya, V. A. *Elektrokhimyia* 1993, *29*, 441.
[100] Gorton, L.; Bremle, G.; Csoeregi, E.; Joensson-Pettersson, G.; Persson, B. *Anal. Chim. Acta* 1991, *249*, 43.
[101] Ruzgas, T.; Gorton, L.; Emneus, J.; Marko-Varga, G. *J. Electroanal. Chem.* 1995, *391*, 41.
[102] Yaropolov, A. I.; Tarasevich, M. R.; Varfolomeyev, S. *Bioelectrochem. Bioenerg.* 1978, *5*, 18.
[103] Zhao, J.; Henkens, R. W.; Stoehauerner, J.; O'Daly, J. P.; Crumbliss, A. L. *J. Electroanal. Chem.* 1992, *327*, 109.
[104] Durliat, H.; Courtiex, A.; Comtat, M. *Bioelectrochem. Bioenerg.* 1989, *22*, 197.
[105] Adeyoju, O.; Iwuoha, E. I.; Smith, M. R. *Anal. Proc.* 1994, *31*, 177.
[106] Yamazaki, I.; Tamura, M.; Nakayama, R. *Mol. Cell. Biochem.* 1981, *40*, 143.
[107] Farhangrazi, Z. S.; Fosett, M. E.; Powers, L. S.; Ellis, W. R. *Biochemistry* 1995, *34*, 2866.
[108] Epton, R.; Hobson, M. E.; Marr, G. *J. Organomet. Chem.* 1978, *149*, 231.
[109] Tatsuma, T.; Okawa, Y.; Watanabe, T. *Anal. Chem.* 1989, *61*, 2352.
[110] Smit, M. H.; Cass, A. E. G. *Anal. Chem.* 1990, *62*, 2429.

[111] Shubert, F.; Saini, S.; Turner, A. P. F. *Anal. Chim. Acta* 1991, *245*, 133.
[112] Frew, J. E.; Harmer, M. A.; Hill, H. A. O.; Libor, S. I. *J. Electroanal. Chem.* 1986, *201*, 1.
[113] Garguillo, M. G.; Huynh, N.; Proctor, A.; Michael, A. C. *Anal. Chem.* 1993, *65*, 523.
[114] Wang, J.; Lin, Y.; Chen, L. *Analyst* 1993, *118*, 277.
[115] Anni, H.; Yonetani, T. In *Degradation of environmental pollutants by micro-organisms and their metalloenzymes*; Siegel, H., Siegel, A., Eds.; Marcel Dekker: New-York, 1992; Vol. 28; pp 219.
[116] Csoeregi, E.; Joensson-Pettersson, G.; Gorton, L. *J. Biotechnol.* 1993, *30*, 315.
[117] Dominguez Sanchez, P.; Tunon Blanco, P.; Fernandez Alvares, J. M.; Smyth, M. R.; O'Kennedy, R. *Electroanalysis* 1990, *2*.
[118] Wollenberg, U.; Bogdanovskaya, V. A.; Bobrin, S.; Scheller, F. W.; Tarasevich, M. R. *Anal. Lett.* 1990, *23*, 1795.
[119] Kulys, J.; Schmid, R. D. *Bioelectrochem. Bioenerg.* 1990, *24*, 305.
[120] Kulys, J.; Bilitewski, U.; Schmid, R. D. *Bioelectrochem. Bioenerg.* 1991, *26*, 277.
[121] Ho, W. O.; Athey, D.; McNeil, C. J.; Hager, H. J.; Evans, G. P.; Mullen, W. H. *J. Electroanal. Chem.* 1993, *351*, 185.
[122] Joensson-Pettersson, G. *Electroanalysis* 1991, *3*, 741.
[123] Wang, J.; Ciszewski, A.; Naser, N. *Electroanalysis* 1992, *4*, 777.
[124] Gorton, L. *Electroanalysis* 1995, *7*, 23.
[125] Popescu, I. C.; Zetterberg, G.; Gorton, L. *Biosens. Bioelectron.* 1995, *10*, 443.
[126] Gorton, L.; Joensson-Pettersson, G.; Csoeregi, E.; Johansson, G.; Dominguez, E.; Marko-Varga, G. *Analyst* 1992, *117*, 1235.
[127] Tatsuma, T.; Gondaira, M.; Watanabe, T. *Anal. Chem.* 1992, *64*, 1183.
[128] Deng, Q.; Dong, S. *J. Electroanal. Chem.* 1994, *377*, 191.
[129] Tatsuma, T.; Ariyama, K.; Oyama, N. *Anal. Chem.* 1995, *67*, 283.
[130] Tatsuma, T.; Ariyama, K.; Oyama, N. *Anal. Chim. Acta* 1996, *318*, 297.
[131] Coche-Guerende, L.; Cosnier, S.; Innocent, C.; Mailley, P. *Anal. Chim. Acta* 1995, *311*, 23.
[132] Vreeke, M.; Maidan, R.; Heller, A. *Anal. Chem.* 1992, *64*, 3084.
[133] Horrock, B. R.; Schmidtke, D.; Heller, A.; Bard, A. J. *Anal. Chem.* 1993, *65*, 3605.
[134] Heller, A. *J. Phys. Chem.* 1992, *96*, 3579.
[135] Gregg, B. A.; Heller, A. *J. Phys. Chem.* 1991, *95*, 5970.
[136] Gregg, B. A.; Heller, A. *J. Phys. Chem.* 1991, *95*, 5976.
[137] Katakis, I.; Heller, A. In *Frontiers in Biosensorics: Fundamental Aspects*; Sheller, F. W., Schubert, F., Fedrowitz, J., Eds.; Bukhausen Verlag: Basel, 1997; pp 227.
[138] Calvo, E. J.; Battaglini, F.; Danilowitcz, C.; Wolosiuk, A.; Otero, M. *Faraday. Discuss.* 2000, *116*, 47.
[139] Decher, G. *Science* 1997, *277*, 1232.
[140] Lvov, Y.; Decher, G.; Sukhorukov, G. *Macromolecules* 1993, *26*, 5396.
[141] Lvov, Y.; Lu, Z.; Schenkman, J. B.; Rusling, J. F. *J. Am. Chem. Soc.* 1998, *120*, 4073.
[142] Kinnear, K. T.; Monbouquette, H. G. *Langmuir* 1993, *9*, 2255.
[143] Lvov, Y.; Ariga, K.; Ichinose, I.; Kunitake, T. *J. Am. Chem. Soc.* 1995, *117*, 6117.
[144] Mizutani, F.; Sato, Y.; Yabuki, S.; Hirata, Y. *Chem. Lett.* 1996, *4*, 251.
[145] Onda, M.; Lvov, Y.; Ariga, K.; Kunitake, T. *Biotechnol. Bioeng.* 1996, *51*, 163.
[146] Kinnear, K. T.; Monbouquette, H. G. *Anal. Chem.* 1997, *69*, 1771.

[147] Hodak, J.; Etchenique, R.; Calvo, E. J.; Singhal, K.; Bartlett, P. N. *Langmuir* 1997, *13*, 2708.

[148] Hou, S.-F.; Yang, K.-S.; Fang, H.-Q.; Chen, H.-Y. *Talanta* 1998, *47*, 561.

[149] Calvo, E. J.; Etchenique, R.; Pietrasanta, L.; Wolosiuk, A. *Anal. Chem.* 2001, *73*, 1161.

[150] Forzani, E. S.; Otero, M.; Perez, M. A.; Lopez-Teijelo, M.; Calvo, E. J. *Langmuir* 2002, *18*, 4020.

[151] Narvaez, A.; Suarez, G.; Popescu, I. C.; Katakis, I.; Dominguez, E. *Biosens. Bioelectron.* 2000, *15*, 43.

[152] Bourdillon, C.; Demaille, C.; Moiroux, J.; Saveant, J. M. *J. Am. Chem. Soc.* 1994, *116*, 10328.

[153] Bartlett, P. N.; Pratt, K. F. E. *J. Electroanal. Chem.* 1995, *397*, 61.

[154] Bourdillon, C.; Demaille, C.; Cueris, J.; Moiroux, J.; Saveant, J.-M. *J. Am. Chem. Soc.* 1993, *115*, 1264.

[155] Calvente, J. J.; Narvaez, A.; Dominguez, E.; Andreu, R. *J. Phys, Chem.* 2003, *107*, 6629.

[156] Bogdanovskaya, V. A.; Fridman, V. A.; Tarasevich, M. R.; Scheller, F. W. *Anal. Lett.* 1994, *27*, 2823.

[157] Dequaire, M.; Limoges, B.; Moiroux, J.; Saveant, J.-M. *J. Am. Chem. Soc.* 2002, *124*, 124.

[158] Tatsuma, T.; Watanabe, T. *Anal. Chem.* 1991, *63*, 1580.

[159] Katakis, I.; Heller, A. *Anal. Chem.* 1992, *64*, 1008.

[160] Rosca, V.; Popescu, I. C. *Electrochem. Comm.* 2002, *4*, 904.

[161] Dunford, H. B.; Stillman, J. S. *Coord. Chem. Rev.* 1976, *19*, 607.

[162] Rosca, V.; Muresan, L.; Popescu, I. C. *Rev. Roum. Chim.*, in press.

[163] Bard, A. J.; Faulkner, L. R. *Electrochemical Methods: Fundamentals and Applications*; Wiley: New York, 1980.

[164] Tatsuma, T.; Ariyama, K.; Oyama, N. *J. Electroanal. Chem.* 1998, *446*, 205.

In: New Developments in Electrochemistry Research
Editor: Magdalena Nunez, pp. 85-111
ISBN: 1-59454-544-8

Chapter 3

USE OF EQCM IN CONJUNCTION WITH VOLTAMMETRY TO STUDY MASS TRANSFER AND CHARGE TRANSFER AT ELECTRODE SURFACES

G. Trejo, Y. Meas, R. Ortega and E. Bahena

Centro de Investigación y Desarrollo Tecnológico en Electroquímica (CIDETEQ), Parque Tecnológico Sanfandila, Pedro Escobedo, A. P. 064. Querétaro C.P. 76700, México

L. Salgado

Universidad Autónoma Metropolitana, Area de Electroquímica. Apdo. Postal 55-534, México D.F. 09340, México

ABSTRACT

This chapter describes the application and basis of the electrochemical quartz crystal microbalance (EQCM). In addition, the piezoelectric effect, frequency-mass rate, and the influences of both viscosity and solution density are discussed. The representation of EQCM data as plots of $d\Delta m / dt$ vs. E, known as massograms, is described. It is shown that mass changes due to Faradaic processes can be distinguished from those due to non-Faradaic processes by analyzing simultaneously recorded massograms and voltammograms. Applications of this method are discussed using several examples, including a study of the adsorption of polyethyleneglycol (PEG) 8000 as well as its co-adsorption with Cl^- ions onto Pt-EQCM electrode, underpotential deposition (UPD) of Zn^{2+} ions onto a Pt-EQCM, and Mn electrodeposition onto a Fe-EQCM. These applications give an appreciation of the power of massograms for the identification of processes that are not evident in conventional representations of EQCM data.

1. Introduction

In the past 15 years, great progress has been made in the capabilities of electrochemical analysis techniques. These improvements have sparked growing interest in modern electrochemical research, motivated by the desire for a more detailed understanding of the structure and composition of the metal–solution interface. Diverse methods have been used to study this interface, including electrochemical scanning tunneling microscopy (EC-STM), electrochemical atomic force microscopy (EC-AFM), electrochemical Raman spectroscopy, and the electrochemical quartz crystal microbalance (EQCM). In the past decade EQCM has emerged as a powerful technique, capable of detecting the small mass changes that accompany electrochemical processes at electrode surfaces.

EQCM is a simple and cost-effective method that gives much useful information for the study of processes at electrode surfaces such as adsorption of species, underpotential deposition (UPD), multilayer deposition and dissolution, and formation of polymer films. Moreover, EQCM data is extremely useful in assigning reaction mechanisms to electrochemical processes.

The utility of EQCM derives from the extraordinary sensitivity with which it can measure mass changes during electrochemical experiments. The data obtained using EQCM, typically simultaneous measurements of mass and current (or charge), enable a more complete analysis of the processes occurring at the electrode–solution interface.

For a quartz crystal with a typical resonance frequency of 5 MHz, EQCM has a theoretical sensitivity of 0.057 Hz cm^2 ng^{-1}, and can typically be operated with a noise level of < 1 Hz; under these conditions the minimum detectable mass is approximately 10 ng cm^{-2} [1]. Higher sensitivities can be achieved by using quartz crystals that operate with fundamental frequencies greater than 5 MHz, where the sensitivity increases with the square of the fundamental frequency [f_0^2].

2. Basic Principals

The quartz crystal microbalance (QCM) has been used for many years as a mass sensor in experiments in vacuum and gas phase. This device was first developed by Sauerbrey in 1959 [2], and in 1981 Nomura et al. [3] adapted it to the measurement of mass in a liquid during electrochemical experiments. The technique of using QCM in situ in electrochemical experiments is known as EQCM or electrochemical quartz crystal nanobalance (EQCN).

EQCM has come to be an important tool in electrochemistry. The basic theory and principles of EQCM have been extensively discussed by various authors, most notably Buttry and Ward [4], Ward [5], and Hepel [6].

This chapter describes the basic principles of the EQCM technique, with the aim of showing the application of EQCM and the relevant information that can be obtained in particular electrochemical systems.

2.1. Piezoelectric Effect

To understand the functioning of the EQCM, it is necessary to understand the piezoelectric effect. In 1880 Jacques and Pierre Curie discovered that application of mechanical tension to the surface of various crystals provokes an electrical potential across the crystal whose magnitude is proportional to the applied tension. This phenomenon, known as the piezoelectric effect, occurs only in materials that are acentric, that is, those that crystallize into noncentrosymmetric space groups.

Subsequent to this initial discovery, the Curies discovered the reverse piezoelectric effect, whereby a crystal can be mechanically deformed by applying a static electric field between its opposing faces. This phenomenon is the basic operational principle of QCM. Moreover, if an oscillating electric field is applied to a quartz crystal at its natural frequency, the crystal resonates, generating a transverse acoustic wave that propagates from one crystal face to the other. By exploiting this effect, it is possible to monitor mass changes at the crystal surface by monitoring variations in the natural frequency of the crystal.

The fundamental frequency [f_0] of the acoustic wave is given by

$$f_0 = \frac{v_{tr}}{2t_Q} \tag{1}$$

where v_{tr} is the transverse velocity of sound in an AT-cut quartz crystal (0.34×10^4 m s^{-1}) and t_Q is the thickness of the resonator.

The quartz crystal commonly used in QCM is AT-cut, where the term AT simply refers to the orientation of the crystal with respect to its long faces. The AT crystal is fabricated by cutting a block of quartz rod at an angle of approximately 35° with respect to the crystallographic axis *x*. AT crystals are widely used because they can access relatively high frequencies, exhibit excellent frequency-temperature stability, and are inexpensive.

In electrochemical applications, the quartz crystal is covered with a metallic film on both faces. One face is exposed to the solution in the electrochemical cell, while the other is exposed to the frequency sensor. The face exposed to the solution operates as the working electrode, and thus the electrochemical perturbation of this electrode can be used to simultaneously measure the electrochemical response and changes in frequency (or mass) at the electrode surface. Electrodes comprised of a quartz crystal coated with films of Ag, Al, Au, Co, Cr, Cu, Fe, Ni, Pt or Zn are produced commercially. The majority of metal-coated quartz crystal electrodes require an adhesion sublayer (20-50 nm), usually made of Cr, Si or Ti.

2.2. Mass–Frequency Relation

The relation between mass change and frequency change is given by the Sauerbrey equation [2]:

$$\Delta f = -2f_0^2\left(\mu_q\rho_q\right)^{\frac{-1}{2}}\Delta m / A \qquad (2)$$

where Δf is the measured frequency change, f_0 is the fundamental frequency of the quartz crystal prior to the mass change, Δm is the change in mass, A is the piezoelectric active area, ρ_q is the density of quartz $\left(\rho_q = 2.648 gcm^{-3}\right)$, and μ_q is the shear modulus ($\mu_q = 2.947x10^{11} gcm^{-1}s^{2}$). The Sauerbrey equation assumes that the acoustic impedance of a quartz crystal with a metallic film is identical to that of the quartz crystal alone. In addition, it assumes that the change in frequency is due to the mass deposited on the surface, and that this mass is homogenously distributed over the entire surface of the piezoelectric crystal. The negative sign in the Sauerbrey equation indicates that the addition of mass to the resonator results in a reduction in its resonance frequency and vice versa. Equation 2 is frequently represented as follows:

$$\Delta f = -C_f \Delta m \qquad (3)$$

where C_f is a proportionality constant, known as the sensitivity factor, which is generally determined by calibration.

2.3. Factors that Contribute to the Frequency Change

In EQCM applications it is important to consider the factors that can contribute to the displacement of the frequency so as to obtain reliable results. In addition to the displacement of the frequency due to mass changes (Δf_{mass}), other factors causing frequency displacement are the solution viscosity ($\Delta f_{viscosity}$), the surface roughness ($\Delta f_{porosity}$), the surface stress at the electrode (Δf_{stress}), compression effects ($\Delta f_{pressure}$), temperature effects ($\Delta f_{temperature}$), as well as viscoelastic effects of the film. Thus the change in the global frequency can be expressed as follows:

$$\Delta f = (\Delta f_{mass}) + (\Delta f_{viscosity}) + (\Delta f_{porosity}) + (\Delta f_{stress}) + (\Delta f_{pressure}) + (\Delta f_{temperature}) \qquad (4)$$

The formation of viscoelastic films increases the complexity of the problem because frequency changes in such films are due to both mass changes and changes in the film viscosity.

For viscoelastic films, Reed at al. [7] derived an equation for the case where both the shear modulus and the viscosity of the film are taken into account:

$$\Delta f = -2f_0^2 \Big/ \left(\rho_q\mu_q\right)^{\frac{1}{2}}\left[\left(\Delta m/A\right) + \left\{\Delta\left(\eta_L\rho_L\right)^{\frac{1}{2}} / \left(4\pi f_0\right)^{\frac{1}{2}}\right\}\right] \qquad (5)$$

The resonance frequency change of the shear vibration of a quartz crystal with piezoelectrically active area A characterized by shear modulus μ_q, density ρ_q, and resonance frequency f_0, in contact with a liquid of density ρ_q and viscosity η_1, is not only affected by rigid mass changes Δm, but also by viscosity changes $\Delta(\eta_1\rho_1)$ in the interface.

The interaction of a piezoelectric electrode with a viscous medium will give rise to additional charge on the electrode and thus to a decrease in the frequency. Kanazawa and Gordon [8] devised an equation relating the frequency change induced by immersion in a solution to the density ρ_L and viscosity η_L of that solution:

$$\Delta f = -f_0^{\frac{3}{2}}\left(\eta_L\rho_L / \pi\mu_q\rho_q\right)^{\frac{1}{2}} \tag{6}$$

In addition, Tsionsky et al. [9] developed a model for determining the film viscosity when there exists an excess of ions at the surface.

Surface roughness can drastically affect the frequency measurements [10-12]. However, provided the roughness is not greater than the wavelength of the acoustic wave, the behavior should conform to the Sauerbrey equation (Eq. 2). When the surface roughness remains constant during an electrochemical experiment, its contribution can be eliminated or at least minimized by calibration of the system.

It has been observed that the resonance frequency of a quartz crystal depends on the elastic energy stored in the quartz. This effect signals the possibility of simultaneous measurements of the changes in mass and surface stress at the electrode [13].

Despite the studies performed to date, many details of EQCM remain poorly understood, principally those mentioned above as well as those associated with the current distribution on the crystal [14]. However, when used appropriately, EQCM gives valuable information that was not previously available.

For applications in which EQCM is used as a sensor of mass changes in thin films on electrodes, the most important factor is the effect of changes in mass, Δf_{mass}, which, as mentioned above, can be separated from the other contributions.

3. Interpretation of Results

EQCM results can be interpreted in a relatively direct manner. Given that the electrochemical charge represents the total number of electrons transferred in a particular electrochemical process, this charge is related to the change in mass that occurs at the electrode surface. Under ideal conditions, the change in frequency measured by EQCM will be proportional to the electrochemical charge and can be related to the apparent molar mass (M_w) via the equation:

$$\Delta f = \frac{M_w C_f Q}{nF} \tag{7}$$

where M_W is the apparent molar mass (g mol^{-1}), Q is the electrochemical charge, n is the number of electrons involved in the electrochemical process, F is the Faraday constant, and C_f (Hz g^{-1} cm^2) is the sensitivity constant (derived from the Sauerbrey equation; Eq. 3). Plots of Δf vs. Q are particularly useful for determining $\frac{M_w}{n}$, which represents the molar mass per transferred electron [15-18]. This calculation requires precise knowledge of C_f, which is obtained by calibration of the EQCM through the electrodeposition of copper or silver [19-20]. Another commonly used format for the analysis of EQCM data are plots of Δm vs. E, which, in conjunction with the corresponding voltammogram, permit analysis of the variation of mass in different zones of the voltammogram.

An alternative way to represent EQCM data proposed recently by Snook et al. [21] is to plot the rate of mass transfer as a function of the potential (i.e., $d\Delta m/dt$ vs. E). From such plots, known as massograms, the rate of mass transfer can be directly compared with the rate of charge transfer (j) at the electrode surface. In this sense a massogram is completely analogous to a voltammogram. An additional advantage of using massograms is that, by comparing massogram and their corresponding voltammogram, it is possible to differentiate between mass changes due to Faradaic and non-Faradaic processes. Fig. 1 shows a typical voltammogram for platinum in 0.1 M $HClO_4$ (curve a, Fig. 1) and its corresponding massogram (curve b, Fig. 1). In Faradaic processes the charge flux is proportional to the mass flux, as can be seen in the j vs E and (dΔm/dt) vs E curves (curves a and b in Fig. 1), associated with the formation and reduction of surface oxides on Pt (region III, Fig. 1). In addition, in region I of the Fig. 1, the mass flux varies in both the positive and negative potential scans due to the adsorption and desorption of water molecules. Faradaic processes that involve gas formation (e.g., evolution of hydrogen and oxygen) are not detected by EQCM and thus the mass flux associated with these processes is null. Additionally, the existence of a mass flux without an associated current flux is evidence of mass transfer due to a non-Faradaic process (e.g., those associated with adsorption of species from solution).

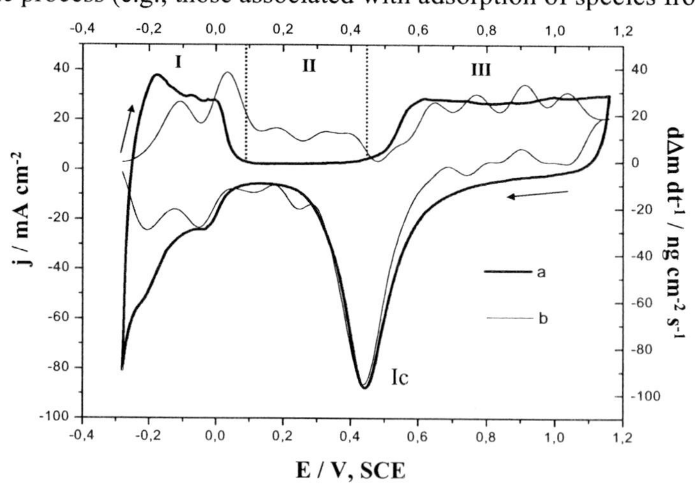

Figure 1. a) Cyclic voltammogram of a Pt crystal in S_0 (0.1 M $HClO_4$), b) Massogram corresponding to the cyclic voltammogram

This chapter outlines the use of massograms through three examples: the adsorption of polyethyleneglycol 8000 (PEG8) and its coadsorption with Cl^- ions; the UPD of Zn^{2+} ions onto Pt in chloride-based acidic media; and the electrodeposition of Mn onto Fe in chloride-based acidic media.

4. Selected Examples of the Application of EQCM and Massograms

4.1. Study of the Adsorption of Polyethyleneglycol 8000 (PEG8) Onto Pt and its Coadsorption with Cl^- Ions in a Perchloric Acid Solution [22][1]

The in situ measurement of the adsorption of organic molecules on an electrode surface is a very important aspect of electrochemistry. EQCM has proved to be extremely useful for the in situ study of adsorption processes [23-29].

The study of the adsorption of polyethoxlyated (PEG) compounds onto metallic surfaces is of great technological importance on account of the widespread use of these compounds as additives in metal electrodeposition processes [30-36] to improve coating quality. The technological importance of these compounds has led to much recent interest in their adsorption behavior. Using transmission electron microscopy (TEM), Hope at al. [37] showed that aggregates of PEG molecules remain hidden during the electrodeposition of copper onto steel. Based on Raman spectroscopy results, Healy et al. [38] proposed that during the electrodeposition of copper, PEG is adsorbed in two different ways: (1) a predominant adsorption close to the open circuit potential, probably due to the adsorption of a complex of copper with PEG as ligand; and (2) another adsorption at more negative potentials that can be associated with the adsorption of the neutral molecule PEG. Similar results were obtained by Bonou et al. [39] using ellipsometry. In spite of the numerous studies that have been made of the influence of PEG and Cl^- on electrodeposition processes, little is known regarding the mechanism by which PEG is adsorbed onto the electrode surface. The objective of this work was to demonstrate the utility of the EQCM technique and massograms to elucidate the adsorption processes of PEG8 onto Pt.

4.1.1. Experimental

The electrochemical and microgravimetric study was carried out in a conventional three-electrode cell with a water jacket. A quartz crystal microbalance (Maxtek Mod. 710) and a potentiostat/galvanostat (EG & G PAR Mod. 273A) controlled by independent computers running the software PM710 and EG & G M270 respectively, were used to simultaneously measure electrochemical parameters and the resonance frequency of the quartz crystal. An AT-cut quartz crystal of nominal frequency f_0 = 5 MHz, covered on both sides with a platinum film (Maxtek, CA), was used as the working electrode. The geometric area of the platinum electrode (Pt-EQCM) was 1.37 cm^2. The roughness factor of this electrode was

[1] Reprinted from Electrochimica Acta, 49, E. Bahena, P. F. Méndez, Y. Meas, R. Ortega, L. Salgado, G. Trejo, "An EQCM study of polyethyleneglycol 8000 adsorption and its coadsorption with Cl- ions on Pt in perchloric acid solutions", Pages 989-997., Copyright (2004), with permission from Elsevier.

approximately 3.3, as estimated from the amount of adsorbed hydrogen at potentials of E > 0.05 V (RHE).

4.1.2. Results and Discussion

The study was carried out by voltammetry in the potential range -0.28 to 1.15 V vs SCE in a 0.1 M $HClO_4$ base solution (S_0) containing a range of PEG8 concentrations ($1x10^{-6}$ to $8x10^{-6}$ M). The potential scan was initiated in the positive direction beginning from E = -0.28V vs SCE. Fig. 2 shows voltammograms of Pt in S_0 with different concentrations of PEG8. In the absence of PEG (i.e., base solution S_0), the j vs E profile has characteristics typical of Pt in $HClO_4$ [18], with three distinct potential regions: region I (-0.28 to 0.1 V) corresponding to the UPD region of hydrogen adsorption-desorption (UPD H); region II (0.1 to 0.45 V) corresponding to the double layer charge, and region III (0.45 to 1.15V) corresponding to formation of the oxide film. On inverting the potential scan direction to the negative direction at 1.15 V, the formation of peak Ic is observed, which is associated with the reduction of the surface oxide.

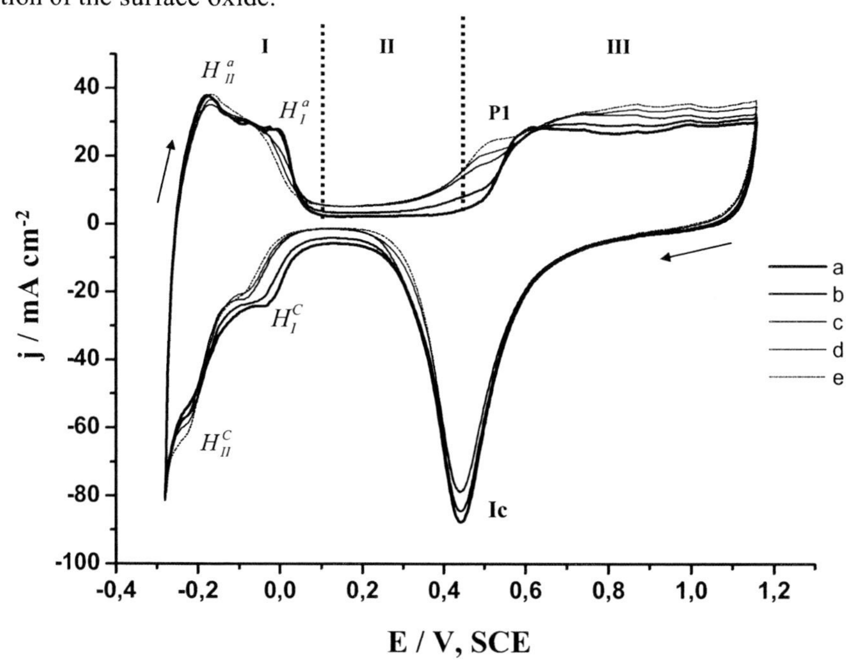

Figure 2. A series of cyclic voltammograms for Pt in: (a) S_0 (0.1 M $HClO_4$), S_0 + different concentrations of PEG8: (b) 10^{-6} M, (c) $4X10^{-6}$ M, (d) $6X10^{-6}$ M. (e) $8X10^{-6}$ M

The addition of PEG8 to the base solution (S_0) causes significant changes in the voltammogram. In region I (Fig. 2), the presence of PEG8 causes a decrease in the current (cathodic and anodic) associated with the adsorption and desorption of hydrogen, principally that which derives from the strongly bound hydrogen (H_I^c and H_I^a, Fig. 2). This effect becomes more evident as the concentration of PEG8 in the solution base is increased. However, the current associated with the UPD adsorption-desorption of hydrogen was not completely abolished, indicating that only a fraction of the active sites for the adsorption of hydrogen are blocked by the adsorbed species [40-41]. The potential interval corresponding to region II in the voltammograms decreases with increasing PEG8 concentration due to the formation of peak P1 at around 0.45 V. The intensity of this peak increases with increasing PEG8 concentration, indicating that it is associated with the oxidation of PEG8. On inverting the potential scan to the negative direction at 1.15 V, a small decrease is observed in the intensity of cathodic peak Ic, suggesting that the additive PEG8 inhibits the formation of surface oxides.

Fig. 3 shows the change in surface mass per unit area (Δm) as a function of the potential (Δm vs E), which was recorded simultaneously with the voltammograms in Fig. 2. The offset of Δm =0 was fixed at E = -0.28 V. Three distinct potential intervals are clearly observed: (I) -0.28 to 0.1 V, corresponding to the potential region of the UPD adsorption-desorption of H; (II) 0.1 to 0.45 V (depending on the concentration of PEG8), corresponding to the region of the double layer charge; and (III) 0.45 to 1.15 V, in which the surface oxide film formation takes place and, where appropriate, the PEG8 oxidation.

During the potential scan in the positive direction recorded in the absence of PEG8 (curve a, Fig. 3), a continuous increase in mass is observed in the potential interval of -0.28 to 1.15 V, with different values of the slope (dΔm/dE) in each potential region. The mass increase in region I has been attributed by various authors to the replacement of adsorbed hydrogen (H_{ads}) by water molecules during the desorption of H_{ads} [40-41]. In the region of the double layer charge (region II), the mass increase is also associated with the continuous adsorption of water molecules and anions [15-18] onto the electrode surface. The mass increase observed in region III is greater than those observed in regions I and II, and is associated with the adsorption of oxygen onto the electrode surface (or surface oxidation).

It is important to observe that, on switching the potential scan to the negative direction starting from 1.15 V, Δm remains almost constant over the potential range 1.15 to 0.6 V, and then rapidly diminishes at potentials less than 0.6 V. This leads to a significant hysteresis, indicative of irreversibility between the processes of formation and reduction of surface oxides on the platinum surface [40-41]. In the potential range 0.15 to 0.25 V, the values of Δm show an appreciable overlap with those obtained in the potential scan in the positive direction. However, the decrease in Δm in the interval of –0.28 to 0.15 V differs from that obtained in the positive scan, giving rise to a small hysteresis (wave H). At the end of the potential scan, Δm returns to its initial value. The fact that the final and initial masses at –0.28 V are the same demonstrates that the mass change is reversible, and that no Pt is dissolved during the potential scan cycle in this medium.

Introduction of PEG8 into the solution causes important changes in the plots of Δm vs E (curves b-d, Fig. 3) with respect to the plot obtained in the absence of this additive. In region I, the variation in Δm during the scan in the positive direction observed in the presence of PEG8 is similar to that observed in the absence of PEG8. In region II, Δm increases to a greater degree in the presence of PEG8 than in the absence of the additive. This greater increase is associated with the oxidation of PEG8 on the Pt surface. In region III, Δm increases rapidly and to a greater degree than in the absence of the additive due to two processes: the formation of a film of PtO on the surface and the oxidation of PEG8.

On switching the potential scan to the negative direction starting from 1.15 V, a hysteresis between the processes of surface oxidation and reduction is observed in the interval 0.5 to 1.15 V, with the magnitude of this hysteresis decreasing with increasing concentration of PEG8 in the solution. At less positive potentials (around 0.3 V), peak A1 forms due to the variation in mass associated with the adsorption-desorption of PEG8. On continuing the potential scan in the negative direction, a hysteresis (waves H) is observed for solution S_0. The intensity of both processes (A1, H) increases with increasing PEG8 concentration. Thus, adsorption of PEG8 increases the irreversibility of the processes that occur in the zone of the double layer and hydrogen adsorption-desorption. It is important to note the existence of an accumulated mass ($\Delta m > 0$) on the Pt surface at the end of the potential scan, which demonstrates that the mass change due to PEG8 adsorption is irreversible.

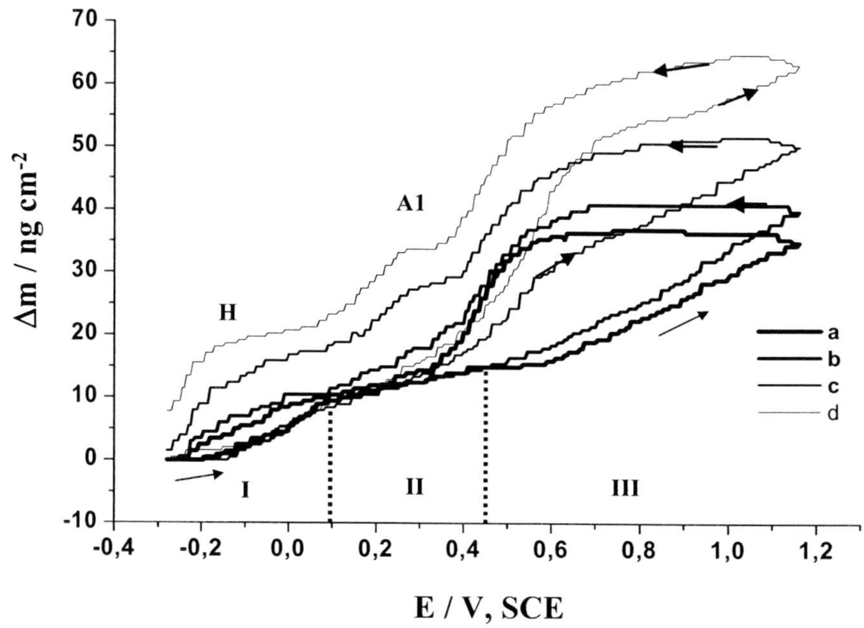

Figure 3. EQCM Δm data plotted against potential; these results were recorded at the same time as the cyclic voltammograms in Fig. 2. (a) S_0, (b) $S_0 + 10^{-6}$ M PEG8, (c) $S_0 + 6X10^{-6}$ M PEG8, (d) $S_0 + 8X10^{-6}$ M PEG8

The peak associated with the adsorption-desorption process of PEG8 (peak A1, Fig. 3) is not observed in the corresponding voltammograms in Fig. 2, suggesting that this process is non-Faradaic. Fig. 4 shows a family of massograms obtained for Pt in solution S_0 with different concentrations of PEG8; these massograms show the mass flux of the voltammograms in Fig. 2. During the potential scan in the positive direction, a peak associated with the oxidation of PEG8 is observed (Peak O1) (positive mass flux); the intensity of this peak increases with increasing concentration of PEG8 in the solution. On switching the potential scan to the negative direction at 1.15 V, a peak is observed associated with the reduction of the Pt surface oxide (peak Ic) (negative mass flux). The intensity of this peak decreases slightly with increasing PEG8 concentration. Subsequently, at less positive potentials, a peak (IcI; 0.25 V) is observed, which is associated with the peak A1 in Fig. 3. Another peak (IcII) is observed at E = -0.2 V, which corresponds to the formation of wave H in Fig. 3. The peaks (IcI and IcII) associated with the two processes (i.e., A1 and H) increase in intensity with increasing concentration of PEG8 in the solution. At the end of the potential scan cycle, the mass flux does not return to its initial value, demonstrating the irreversibility of the adsorption process. Comparison of the voltammograms and massograms clearly reveals that the partial adsorption-desorption process of PEG8 is non-Faradaic, a conclusion that cannot be made based on the voltammetry data alone.

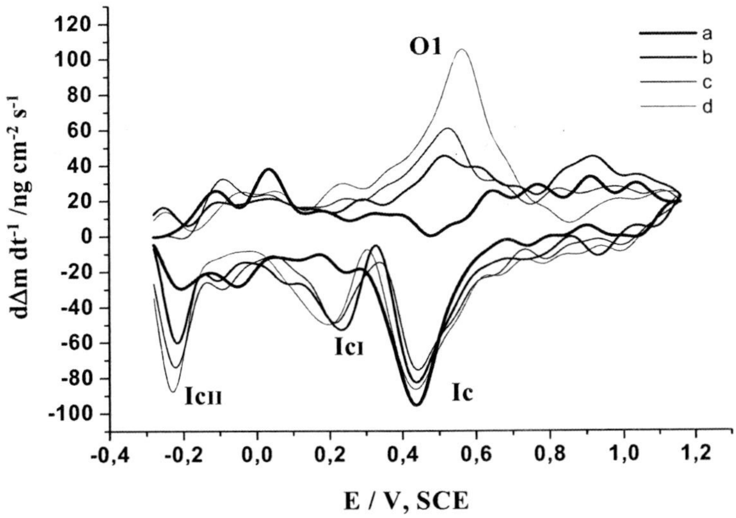

Figure 4. A series of massograms obtained at the same time as the cyclic voltammograms in Fig. 2 (a) S_0, (b) $S_0 + 10^{-6}$ M PEG8, (c) $S_0 + 6X10^{-6}$ M PEG8, (d) $S_0 + 8X10^{-6}$ M PEG8.

To investigate the adsorption of PEG8 during the negative potential scan, a linear potential scan beginning from the resting potential ($E_{rest} = 0.7$ V) was carried out up to a final potential of $E_f = -0.28$ V. The resulting Δm vs E plots are shown in Fig. 5. In this case, the offset of $\Delta m = 0$ was fixed at E_{rest}. The formation of peak A1 and wave H is observed, with the intensities of both of these features increasing with increasing concentration of PEG8. At the end of the potential scan there exists an accumulated mass whose magnitude increases with increasing concentration of PEG8 in the solution. This demonstrates that the PEG8 additive undergoes an partial adsorption-desorption process during the potential scan in the negative direction, causing a fraction of the substrate to remain covered with adsorbed PEG8 ($PEG8_{ads}$). The maximum fraction of PEG_{ads} for the range of concentrations studied was 0.21, which corresponds to 12.10×10^{11} molecules cm^{-2}.

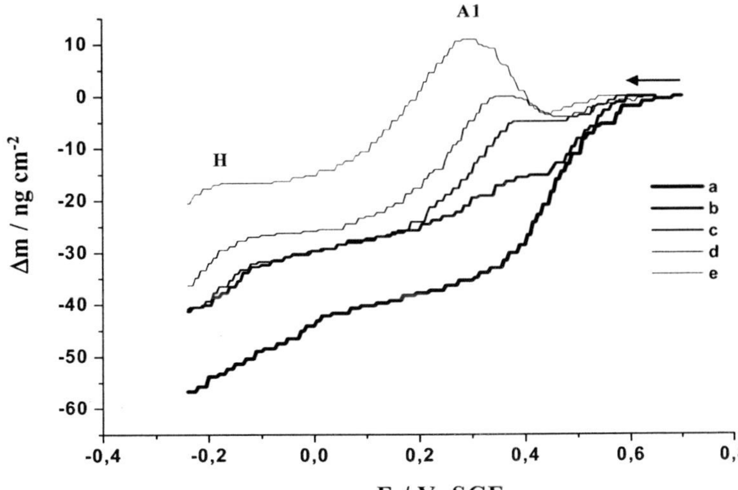

Figure 5. Plots of Δm vs E at different concentrations of PEG8, obtained from linear potential scans. Initial potential (E_{rest}) = 0.7 V, Final potential (E_f) = -0.28 V, scan velocity 50 mV/s. (a) S_0, (b) $S_0 + 10^{-6}$ M PEG8, (c) $S_0 + 4X10^{-6}$ M PEG8, (d) $S_0 + 6X10^{-6}$ M PEG8, (e) $S_0 + 8X10^{-6}$ M PEG8.

4.1.2.1. Study of the Coadsorption of PEG8 and Cl^- on Pt

Fig. 6 shows the voltammograms obtained for Pt in several different solutions: S_0 (0.1 M $HClO_4$), S_1 (S_0 + 0.048 g L^{-1} PEG8), and S_1 with different concentrations of Cl^- (10^{-6} M $\leq$ [KCl] $\leq 10^{-2}$ M). The voltammograms obtained for S_0 and S_1 (curves a and b in Fig. 6) have been described in detail in the previous section. The presence of Cl^- ions in solution S_1 produces significant changes in the j vs E curves compared to the curves obtained for S_0 and S_1; specifically, as the concentration of Cl^- is increased, the potential interval corresponding to the double layer charge widens. This effect is characteristic of the adsorption of Cl^- [40-41], and is associated with the progressive blocking of the initial states (up to 0.85 V) of the developed oxide film that is formed as PtO [41-44]. In addition, increasing the chloride ion concentration causes both a decrease in the intensity of the anodic peak associated with the oxidation of PEG8 (peak P1), and the displacement of this peak to more positive potentials until it completely disappears. This behavior could be due to the Cl^- ions replacing the $PEG8_{ads}$ on the Pt surface, a process that would cause the quantity of surface oxides formed during the potential scan in the positive direction to decrease with increasing concentration of Cl^- ions. This mechanism is supported by the decrease observed in the intensity of the peak associated with the reduction of surface oxides (peak Ic). In the region of UPD adsorption-desorption of hydrogen, the presence of chloride ions causes the hydrogen adsorption-desorption process the be displaced to less positive potentials, which is a characteristic effect of the adsorption of Cl^- on Pt surfaces [40-41].

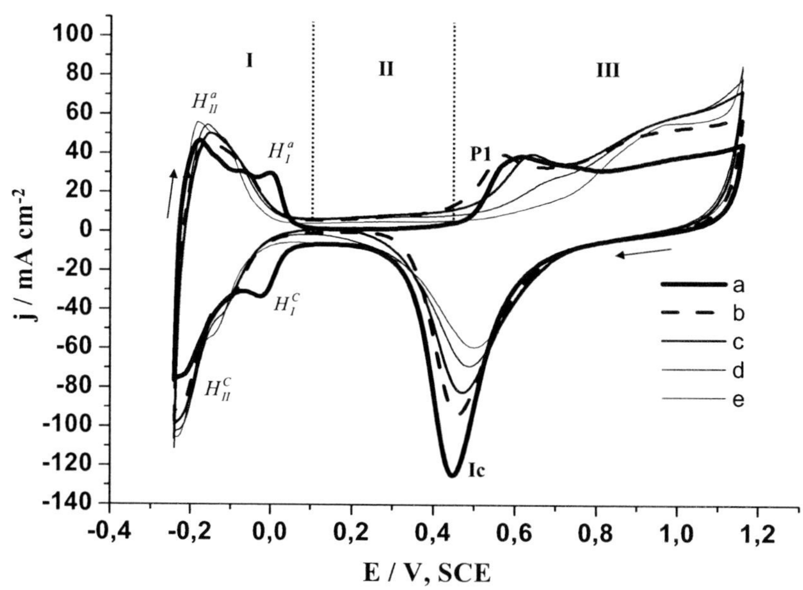

Figure 6. Cyclic voltammograms of Pt in various solutions: (a) S_0, (b) S_1 (S_0 + 6X10^{-6} M PEG8), (c) S_1 + 10^{-5} M Cl^-, (d) S_1 + 10^{-4} M Cl^-, (e) S_1 + 10^{-3} M Cl^-

Fig. 7 shows the massograms obtained for solutions S_0 and S_1 with different concentrations of Cl^-. The presence of Cl^- causes important changes in the massograms. Specifically, addition of Cl^- at concentrations of up to 10^{-2} M causes the peak associated with the oxidation of PEG8 (peak O1) to decrease in intensity and to be displaced toward more positive potentials (behavior similar to that observed by voltammetry, Fig. 6). At more positive potentials (> 1.0 V), the mass flux rapidly decreases down to negative values. This behavior can be explained by considering that, in this potential zone, the evolution of chlorine gas occurs at the Pt surface, which may cause detachment of the adsorbed PEG8. On switching the potential scan to the negative direction, formation of peak Ic, the peak associated with the reduction of the surface oxide, is observed. At less positive potentials (E = -0.25 V), peak IcII is observed, which is associated with the mass flux that occurs during the adsorption of hydrogen onto Pt (desorption of water molecules); the intensity of this peak is independent of the Cl^- ion concentration. When the concentration of Cl^- was 10^{-2} M (curve d, Fig. 7), the behavior observed in the massograms is typical of Pt in a Cl^- solution; specifically, peak O1 disappears due to the displacement of $PEG8_{ads}$ on the Pt surface by Cl^- ions. This result is corroborated by the fact that peak IcI (associated with the adsorption of PEG8) observed for S_1 (curve b, Fig. 7), disappears during the potential scan in the negative direction.

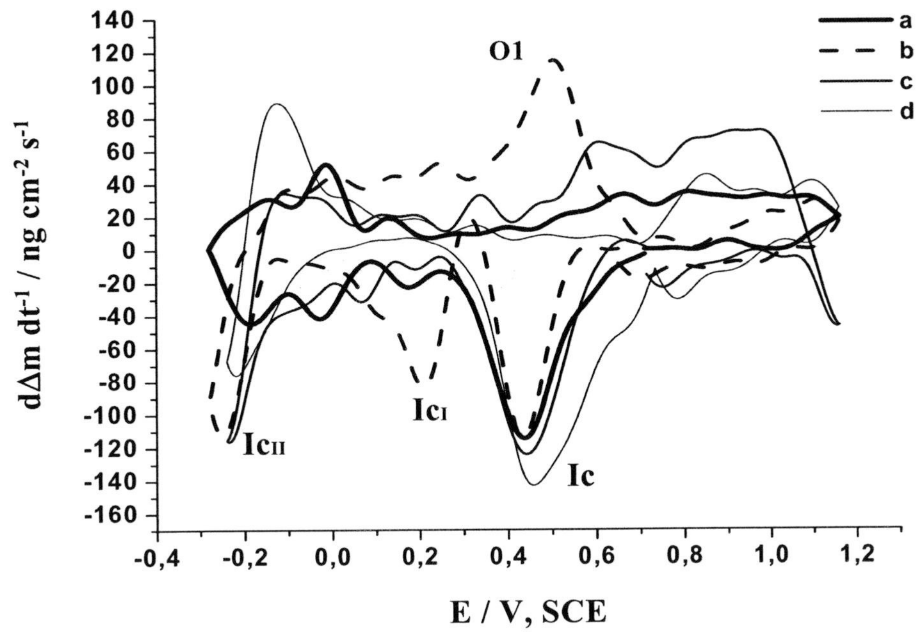

Figure 7. A series of massograms obtained at the same time as the cyclic voltammograms in Fig. 6 (a) S_0, (b) S_1 (=S_0 + 6X10^{-6} M PEG8), (c) S_1 + 10^{-3} M Cl^-, (d) S_1 + 10^{-2} M Cl^-

To investigate the inhibition of PEG8 adsorption by chloride ions, Δm vs E plots were constructed using EQCM measurements. The tests were carried out using a linear potential scan in the negative direction from the resting potential (E_{rest} = 0.7 V) to a final potential of E_f = –0.28 V. Fig. 8 shows the plots obtained for solutions S_0, S_1, and S_1 with different chloride concentrations. In these experiments the offset of Δm = 0 was placed on E_{rest}.

As the concentration of Cl^- ions in solution S_1 was increased, wave A1 gradually disappeared, thus confirming that Cl^- ions inhibit the adsorption of PEG8 during the negative scan.

At the end potential scan for solution S_1, a final accumulated mass (Δm) is observed. The difference between this mass gain and that obtained in solution S_0 corresponds to the mass of adsorbed PEG8. As the concentration of Cl^- ions in S_1 is increased, the final accumulated mass diminishes until, at Cl^-concentrations greater than $8X10^{-6}$ M, the accumulated mass is less than that obtained in solution S_0. This loss of mass with respect to solution S_0 can be attributed to the oxidation of Pt in chloride media [42-43]. Based on these results, it is possible to propose that at sufficiently high concentrations of Cl^-, the PEG_{ads} is displaced from the electrode surface by Cl^- ions under the conditions used in the present study.

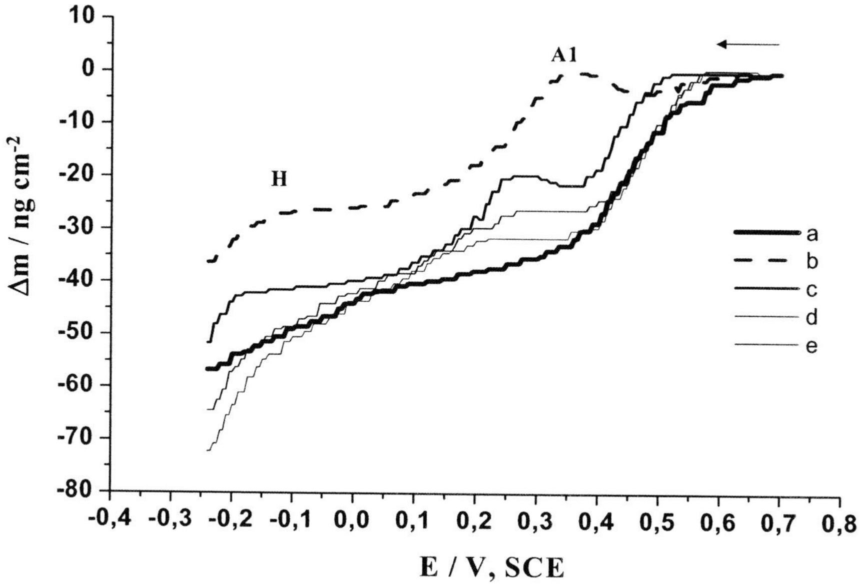

Figure 8. Plots of Δm vs E obtained from linear potential scans. $E_{rest} = 0.7$ V, $E_f = -0.28$ V, v = 50 mV/s, for various solutions: (a) S_0, (b) S_1 (S_0 + $6X10^{-6}$ M PEG8), (c) S_1 + 10^{-5} M Cl^-, (d) S_1 + 10^{-4} M Cl^-, (e) S_1 + 10^{-3} M Cl^-

It is clear from the present results that the joint use of massograms and voltammograms makes it possible to differentiate between mass changes due to Faradaic and non-Faradaic processes. In the present study the joint use of these two techniques revealed that the process of PEG8 adsorption is non-Faradaic and strongly depends on the applied potential, with partial adsorption-desorption of PEG8 being observed during the cathodic potential scan. Moreover, this methodology clearly showed that in a medium with sufficiently high concentrations of Cl^- (10^{-3} M), the chloride ions displace the PEG_{ads} on the Pt surface.

4.2. EQCM Study of the Underpotential Deposition of Zn^{2+} Ions onto Pt in Chloride-Based Acidic Media

Another application of the EQCM technique is to study the electrodeposition of metals at underpotentials (UPD) [45-46], that is, at more positive potentials than the thermodynamic potential of the equilibrium M^{n+}/M. UPD processes are of great theoretical and practical importance in various areas of electrochemistry, including electrocrystallization, electrocoating and electrocatalysis. In this sense the UPD of Zn^{2+} ions has a great impact on our knowledge of the electrodeposition of zinc alloys, given that the UPD of Zn^{2+} ions plays a fundamental role in the anomalous codeposition mechanism [47] during the formation of zinc alloys with iron-group metals (e.g., Zn-Fe, Zn-Ni, Zn-Co).

The UPD of Zn^{2+} ions has been studied by Aramata et al. [48-50] on poly- and monocrystalline Pt, Pd and Au substrates. They investigated the effect of pH on the anodic displacement of the peak corresponding to the dissolution of Zn deposited by UPD, and suggested a process involving the transfer of two electrons in Pt substrates. In addition, it has been reported that the UPD of Zn^{2+} ions is sensitive to the crystallographic orientation of the surface, with the (110) orientation being the most favorable [50].

The influence of the anions on the UPD of Zn^{2+} has been studied by various authors [51-54]. Experiments on the adsorption of Cl^-, HSO_4^- (SO_4^{2-}) and $H_2PO_4^-$ onto Pt surfaces induced by the coadsorption of Zn adatoms showed that the specific adsorption of HSO_4^- (SO_4^{2-}) is significantly greater than the adsorption of Cl^- onto the same Zn adlayer. The model proposed to explain this process involves three steps: 1) desorption of anions from the substrate; II) UPD of Zn^{2+} ions; and III) adsorption of anions onto Zn deposited by UPD. This model is supported by results obtained using a variety of techniques, including voltammetry, radiotracer, FTIR and EQCM. In a study using voltammetry and rotating ring-disc voltammetry, Mascaro et al. [55] showed that the coadsorption of HSO_4^-, ClO_4^- and F^- onto the substrate or onto the adatoms influenced the nature of the Zn adlayer.

4.2.1. Experimental

An AT-cut quartz crystal of nominal frequency $f_0 = 5$ MHz, covered on both sides with a platinum film (Maxtek, CA), was used as the working electrode. The geometric area of the platinum electrode (Pt-EQCM) was 1.37 cm^2. The roughness factor of this electrode was approximately 3.3, as estimated from the amount of adsorbed hydrogen at potentials of E > 0.05 V (RHE).

4.2.2. Results and Discussion

Fig. 9 shows the cyclic voltammograms (CV) of Pt in the potential range -0.185 to 1.285 V vs SCE, obtained from three different solutions: base solution S_0 (0.5 M H_2SO_4) (curve a); solution S_1 (S_o + 10^{-3} M Zn^{2+}) (curve b) and solution S_2 (S_1 + 10^{-2} M KCl) (curve c), recorded at 50 mV s^{-1}.

Comparison of the CVs for solutions S_0 (curve a, Fig. 9) and S_1 (curve b, Fig. 9) reveals the effect of the Zn^{2+} ions on the UPD H process. In the presence of Zn^{2+} (curve b), the anodic and cathodic current densities increase when E < 0.165V (UPD H, region I); three anodic peaks are observed [peak Ia (-0.060 V), peak IIa (0.000 V), and peak IIIa (0.048 V)] as well as the corresponding cathodic peaks [peak Ic (-0.096 V), peak IIc (-0.044 V), and peak IIIc (0.032V)]. An increase in the intensities of the peaks H_{D1} and H_{A1} and a decrease in H_{D2} and H_{A2} are observed, an effect that can be attributed to UPD of Zn^{2+} ions.

The cathodic charge density (Q^c_{upd}) obtained by integrating the plot of j vs E in this potential range (-0.185 to 0.165 V) was 0.190 mC cm^{-2}. The charge density observed in this region was 17% higher than that obtained in the absence of Zn^{2+} ions.

In addition, the presence of Zn^{2+} ions has no significant influence on the processes occurring in the electrical double layer charge region (region II), nor on the formation and reduction of surface oxides (region III). Similar results were reported by Aramata et al. [48-

49], associating the changes observed in region I of the voltammograms with the UPD of Zn^{2+} ions and dissolution of Zn deposited by UPD.

Addition of Cl^- to the solution (i.e., solution S_2) causes an important change in the voltammograms (curve c, Fig. 9). The voltammogram obtained in the presence of Cl^- shows the following features. In region I, only two cathodic peaks (Ic and IIc) and two anodic peaks (Ia and IIa) are observed, suggesting that the Cl^- ions inhibit the adsorption of the species associated with peak III_c and promote the adsorption of the species associated with peaks I_c and II_c in curve b of Fig. 9. In addition, the current density in the region I is greater (2%) than that observed in the absence of Cl^- ions (curve b). Also, the potential interval corresponding to the double layer charge (region II) widens, a characteristic effect of the adsorption of Cl^- [41-42] that is associated with the progressive blocking of the initial states (up to 0.7 V) of the formation of the oxide film formed as PtO. This is corroborated by the decrease (23%) observed in the intensity of the peak corresponding to the reduction of the surface oxide (peak O_{red}).

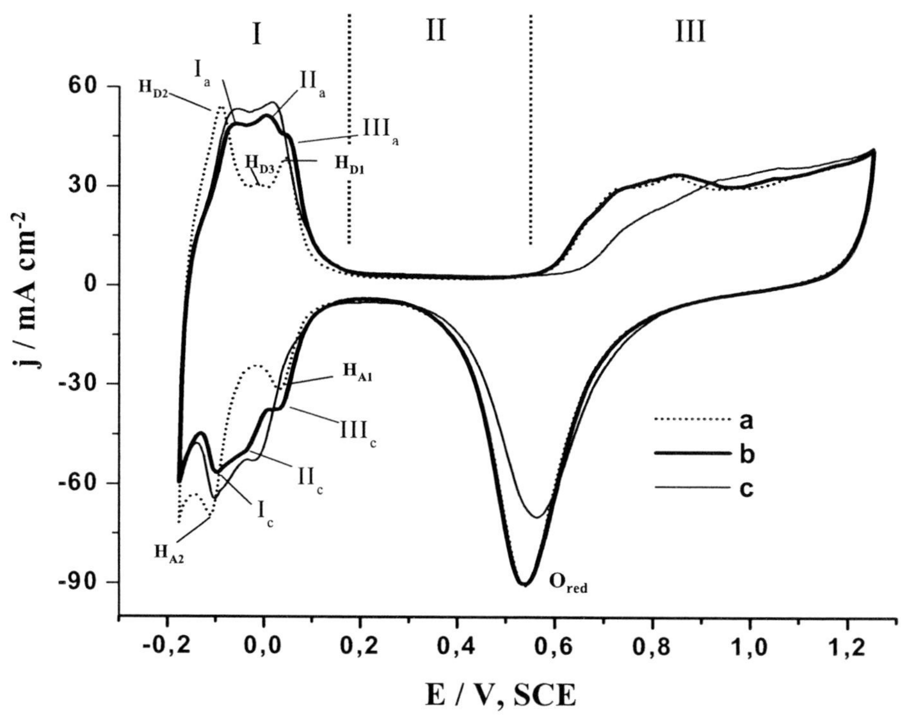

Figure 9. Cyclic voltammograms obtained on Pt in different solutions: a) solution S_0 (= H_2SO_4 0.5 M). b) solution S_1 (= S_0 + 1 x 10^{-3} M Zn^{2+}). c) solution S_2 (=S_1+ 10^{-2} M KCl). Potential scan rate 50 mV s^{-1}.

Fig. 10 shows the profiles of the variation of mass with changing potential (Δm vs E) corresponding to the cyclic voltammograms in Fig. 9; the curves a, b and c in Fig. 10 have been grouped with offset Δm = 0 fixed at the upper potential limit, E = 1.285 V. Curve a shows the change in mass as a function of potential for Pt in solution S_0; the behavior of this system is similar to that discussed in the previous section for the Pt/$HClO_4$ system and is presented in Fig. 10 for comparison purposes.

Comparison of the curves obtained for solutions S_0 and S_1 (curves a and b respectively, Fig. 10) reveals that the changes in mass in regions II and III do not depend on the presence of Zn^{2+} ions in the solution. In the region I, however, the presence of Zn^{2+} ions (curve b, Fig. 10) leads to an increase in mass ($\Delta m_{dep} = 25.1$ ng cm^{-2}) during the potential scan in the negative direction. In addition, switching the potential scan to the positive direction starting from -0.185 V produces a decrease in mass of the same magnitude, suggesting that the UPD adsorption-desorption process of Zn is reversible in this potential range.

It is important to note that the change in mass determined experimentally (Δm_{dep} = 25.1 ng cm^{-2}) during the cathodic scan depends on two factors: mass gain due to the UPD of Zn^{2+} (Δm_{upd}^{Zn}) and mass loss due to the desorption of water molecules ($\Delta m_{ads}^{H_2O}$ = -17.1 ng cm^{-2}, curve a, Fig. 9) that occurs during the adsorption of hydrogen onto the Pt surface. Given that the EQCM does not detect mass changes due to hydrogen adsorption (H_{ads}), it is possible to determine Δm_{upd}^{Zn} using the following equation:

$$\Delta m_{dep} = \Delta m_{upd}^{Zn} + \Delta m_{ads}^{H_2O} \tag{8}$$

where Δm_{dep} is the net increase in mass recorded by EQCM (25.1 ng cm^{-2}), $\Delta m_{ads}^{H_2O}$ is the mass of water molecules desorbed (-17.1 ng cm^{-2}), and Δm_{upd}^{Zn} is the total mass of zinc deposited by UPD. The value of Δm_{upd}^{Zn} obtained was 42.2 ng cm^{-2} (6.45 x 10^{-10} mol cm^{-2} atoms of Zn deposited by UPD). The electric charge density associated with the total mass deposited during the process UPD of Zn^{2+}, Q_{upd}^{Zn} = 0.124 mC cm^{-2}, is less than that obtained by voltammetry for the same process (Q_{upd}^{c} = 0.190 mC cm^{-2}). The excess electric charge is associated with the adsorption of hydrogen (H_{ads}) onto Pt ($Q_{upd}^{H} = Q_{upd}^{c} - Q_{upd}^{Zn} = 0.065$ mC cm^{-2}).

From the EQCM results, the degree of coating by Zn, θ_{Zn}, can be determined using the following equation:

$$\theta_{Zn} = (\frac{Q_{upd}^{Zn}}{2Q_{H,s}}) = 0.29 \tag{9}$$

where Q_{upd}^{Zn} is the cathodic charge density associated with the adsorption of Zn adatoms, and $Q_{H,s}$ is the charge associated with the adsorption of a monolayer of H onto polycrystalline Pt and a transfer of two electrons from Zn [48].

Curve c in Fig. 10 shows the effect of adding Cl^- ions (solution S_2) on UPD of Zn^{2+}. The changes in mass produced in potential regions II and III of the Δm vs E curve (curve c, Fig. 3) are similar to those reported by Zolfaghari at al. [41] for Pt in a 0.05 M H_2SO_4 solution containing Cl^- ions. Specifically, in the double layer region (region II), the presence of Cl^- ions induces a hysteresis between the potential scans in the positive and negative directions; this behavior is attributed to transitory desorption of Cl^- due to the reduction of coadsorbed O species generated during the anodic scan. At more positive potentials (region III), the adsorbed Cl^- ions inhibit the formation of surface oxides during the positive potential scan; thus the mass changes due to the formation and reduction of surface oxides are less than those obtained in the absence of Cl^- ions.

In region I during the potential scan in the negative direction, the Δm vs E curve shows an increase in mass of 26.7 ng cm^{-2} in the interval of 0.109 to -0.185 V, which is 56 mV more positive than in the absence of Cl^- ions. This implies that the presence of Cl^- ions induces a mass increase 6% greater than that recorded in the absence of Cl^-. However, on switching the potential scan to the positive direction beginning from -0.185 V, a loss of mass of 25.1 ng cm^{-2} is observed in the same potential interval. This value is similar to that obtained in the absence of Cl^- (curve b, Fig. 10), suggesting that the quantity of Zn adatoms does not depend on the Cl^- ions in solution. The mass of adsorbed species that are not desorbed (1.6 ng) during the potential scan in the positive direction may, as suggested by Horanyi et al. [52-53], be associated with chloride ions being adsorbed onto the electrode surface and then remaining on the electrode surface [42]. This phenomenon would account for the hysteresis in the UPD of Zn, as is observed in the scan in the positive direction (curve c, Fig. 10). Thus, the increase in charge observed in the voltammogram recorded in the presence of chloride ions (curve c, Fig. 9) corresponds to the increase in the amount of H_{ads} and not to the Zn adatoms.

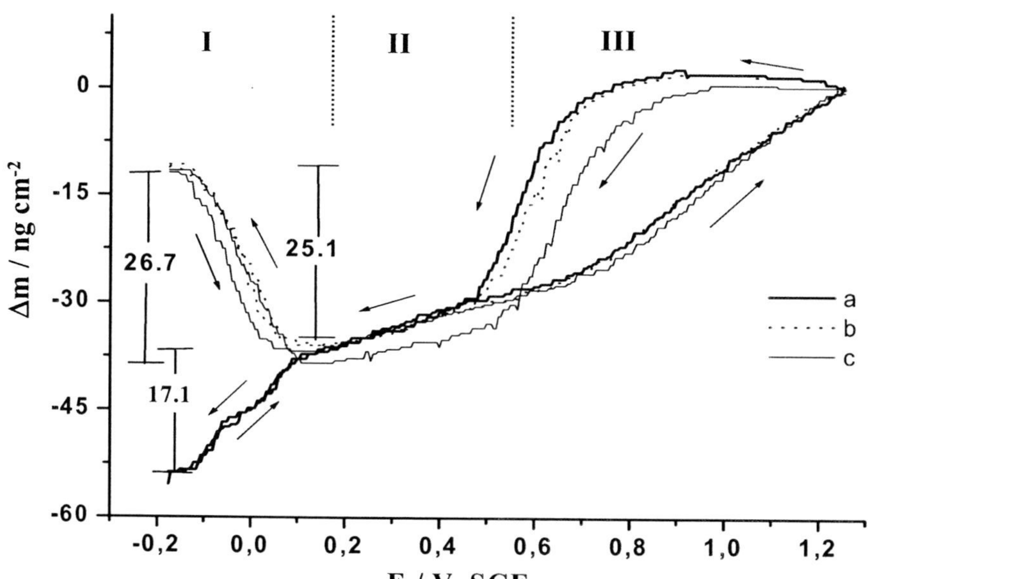

Figure 10. Δm-E curves corresponding to the cyclic voltammograms in Fig. 9. a) solution S_0 (= 0.5 M H_2SO_4). b) solution S_1 (= S_0 + 10^{-3} M Zn^{2+}). c) solution S_2 (=S_1+ 10^{-2} M KCl). Potential scan rate 50 mV s^{-1}.

Figure 11 shows the massograms corresponding to the voltammograms in Fig. 9 for solutions S_0 (curve a), S_1 (curve b), and S_2 (curve c).

In the absence of Zn^{2+} and Cl^- ions (curve a, Fig. 11), the profile of the massogram is characteristic of Pt in 0.5 M H_2SO_4. During the potential scan in the negative direction starting from 1.285 V (curve a, thick line), various peaks of negative mass flux are observed that correspond to loss of mass from the Pt surface. Peak Ic is observed at E = 0.6 V and corresponds to the reduction of surface oxide. Subsequently, at less positive potentials (region

I), two peaks (H_{A1}, H_{A2}) are observed that correspond to the desorption of water molecules (adsorption of hydrogen (H_{ads})) from the Pt surface. On switching the potential scan to the positive direction starting from -0.185 V (curve a, thin line), a positive mass flux (mass gain) is observed in region I of the massogram. This mass flux corresponds to the adsorption of water molecules onto the active sites vacated by the desorption of H_{ads} from the Pt surface; the peaks H_{D1} and H_{D2} are the anodic equivalents of H_{A1} and H_{A2}, while H_{D3} corresponds to the desorption of H_{ads} from different active sites. Additionally, in region III positive mass flux variations are observed due to oxide formation on the surface. The behavior observed in the massogram corresponds to that recorded in the voltammogram (curve a, Fig. 9).

The presence of Zn^{2+} ions (curve b) causes important changes in the massograms. During the potential scan in the negative direction starting from 1.285 V (curve b, thick line), peak Ic is observed, which corresponds to the reduction of surface oxide. Subsequently, in region I, a transition from negative mass flux to positive mass flux (mass gain) is observed, originating from the UPD of Zn^{2+} ions. Two peaks (m1 and m2) appear, with potential values Em_1 (= 0.04 V) and Em_2 (= -0.075 V), which are similar to those of peaks H_{A1} (E_{HA1} = 0.05 V) and H_{A2} (E_{HA2} = -0.075 V), indicating that the UPD of Zn^{2+} ions occurs principally on these sites (H_{A1} and H_{A2}) and with a different rate of mass transfer. Given that the EQCM technique does not detect mass changes due to hydrogen adsorption, the positive mass flux observed corresponds to the UPD of Zn^{2+} ions.

On switching the potential scan to the positive direction starting from -0.185 V (curve b, thin line) a shoulder (S) is observed prior to the peak corresponding to the dissolution of the Zn deposited by UPD (peak D). It is important to observe that the potential values of both the shoulder S (E_S = -0.075 V) and peak H_{D2} (E_{HD2} = -0.075 V) are equal, while the potential of the dissolution peak D (E_D = -0.025 V) is close to E_{HD3} (= -0.018 V). This result suggests that the UPD of Zn^{2+} ions that occurs at the H_{A1} sites undergoes a readjustment on the Pt surface in order to occupy the H_{D3} that have been vacated by desorption.

Another important point is that at potentials more positive than E_D, no other process involving the dissolution of Zn deposited by UPD is observed. Thus the current associated with peak IIIa of the voltammogram (curve b, Fig. 9) corresponds to the desorption of H_{ads}. In addition, in regions II and III, no important changes are observed in the massograms.

Curve c in Fig. 11 shows the massogram recorded in the presence of Cl^- ions. During the potential scan in the negative direction starting from 1.285 V (curve c, thick line), a decrease in the intensity of peak Ic is observed compared to that recorded in the absence of Cl^- ions; this behavior is associated with the decrease in the quantity of surface oxide formed due to the adsorption of Cl^- ions onto the Pt. At less positive potentials (region I), the presence of the Cl^- ions slightly inhibits the Zn UPD process giving rise to peak m1, while favoring the process giving rise to peak m2. In addition, a new process is observed, manifesting as peak M. Thus, three rates of formation are observed for the process UPD of Zn^{2+} ions in the presence of Cl^- ions. When the potential scan is switched to the positive direction starting from -0.185V, a small displacement in the positive direction is observed in the peak corresponding to the dissolution of Zn deposited by UPD (curve c, thin line).

In addition, the potential interval of the double layer region (zone II) widens slightly due to the adsorption of Cl^- ions and to the partial inhibition of the formation of surface oxide, which accounts for the behavior observed in peak Ic.

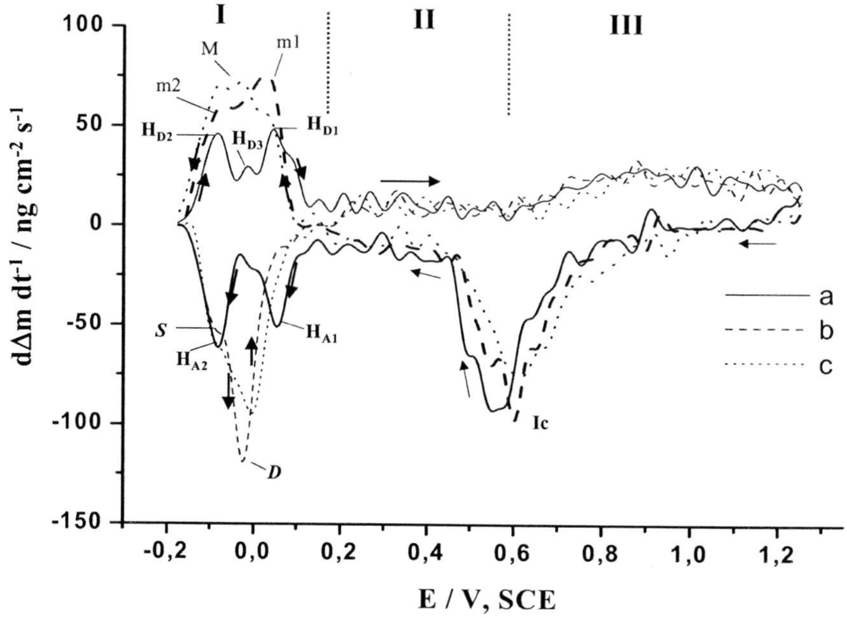

Figure 11. Series of massograms obtained at the same time as the voltammograms in Fig. 9. a) solution S_0 (= 0.5 M H_2SO_4). b) solution S_1 (= S_0 + 1 x 10^{-3} M Zn^{2+}). c) solution S_2 (=S_1+ 10^{-2} M KCl). Potential scan rate 50 mV s^{-1}.

Thus, the joint use of EQCM, massograms and voltammograms makes it possible to differentiate the charge associated with the UPD of Zn^{2+} and the dissolution of Zn deposited by UPD, from the charge associated with the UPD of H. In addition, it was possible to determine that Cl^- ions do not affect the UPD of Zn^{2+} ions in an important way, although they do interfere with the formation of surface oxides of Pt (PtO).

4.3. Electrodeposition of Manganese on Fe In Chloride-Based Acidic Media

Lastly we consider a system characterized by Faradaic deposition with interference from hydrogen evolution. Such behavior is common during the electrodeposition in acidic solutions of metals whose potential for the pair M^{n+}/M is quite negative. When hydrogen evolution occurs, it is generally not possible to discern the electrodeposition process in the voltammogram due to interference from the hydrogen evolution reaction:

$$2H^+_{(aq)} + 2e^- \Leftrightarrow H_{2(g)} \tag{10}$$

In the case of manganese (Mn), the potential of the pair Mn^{2+}/Mn has a value of -1.18 V/SHE. Thus the deposition of Mn occurs simultaneously with hydrogen evolution, causing the voltammograms for this system to have characteristics that are not typical of electrodeposition processes.

Mn coatings or Mn alloys (e.g., Zn-Mn) obtained by electrodeposition are potentially useful as sacrificial coatings for protecting ferrous substrates against corrosion [56-61]. In spite of the potential utility of these coatings, little research has been conducted into the electrodeposition mechanism of metallic Mn. This paucity of studies can be attributed to the difficulty of studying this electrodeposition process due to interference from hydrogen evolution during the reduction process. However the EQCM technique, particularly massograms, offer an alternative method for studying the deposition of Mn. Given that EQCM does not detect mass changes due to the gas formation ($H_{2(g)}$) [21], it is possible to clearly distinguish the mass flux due to the reduction of Mn at the electrode surface.

4.3.1.Experimental

The study was carried out in a base solution (S_0) with the following composition: 2.8 M KCl + 0.4 M H_3BO_3 + 0.6 M $MnCl_2$, pH = 5.0, in the potential range -1.8 to -0.7 V vs SCE. The potential scan was initiated in the cathodic direction starting from the resting potential (E_{rest}) using a Cr/Au/Fe (Fe-EQCM) electrode of resonance frequency 5 MHz (Maxtek).

4.3.2. Results and Discussion

Fig. 12 shows a typical voltammogram obtained for the base solution (S_0) on the Fe-EQCM electrode. During the potential scan in the negative direction a reduction wave (A) is observed in the potential range -1.6 to -0.9 V. Subsequently, at more negative potentials, in the range -1.8 to -1.6 V, a rapid increase occurs in the cathodic current density. The total charge obtained during the scan in the negative direction (Q^c) was 110.07 mC cm^{-2}. On switching the potential scan to the positive direction, starting from -1.8 V, a hysteresis appears between the positive and negative potential scans, indicating the formation of a new phase on the surface of the Fe-EQCM. It is important to observe that over the potential range studied, no oxidation current was recorded. Sylla et al. [61] observed similar behavior during the electrodeposition of Mn on steel, although they did not attempt to explain this observation.

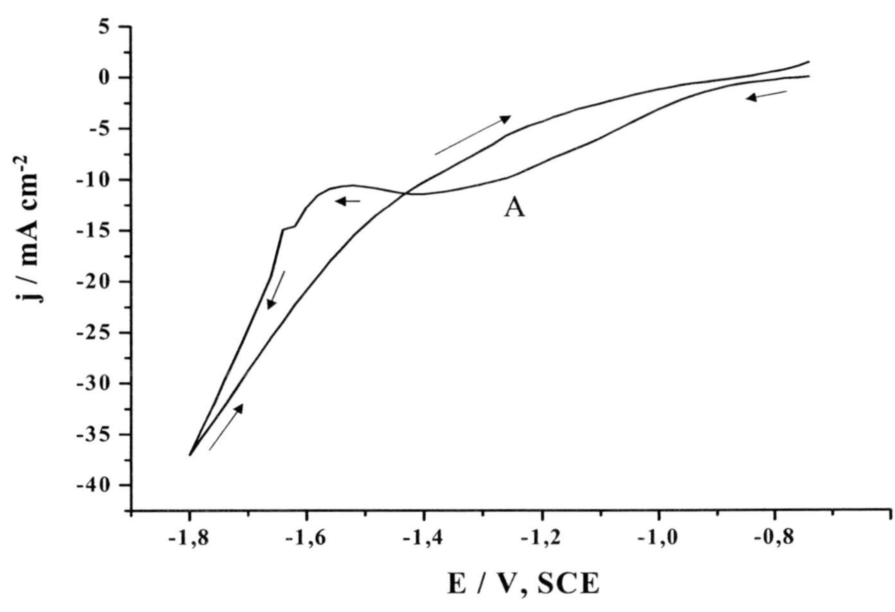

Figure 12. Typical voltammogram obtained on Fe-EQCM electrode in a solution of composition 2.8 M KCl + 0.4 M H_3BO_3 + 0.6 M $MnCl_2$, pH = 5.0.

A possible explanation for this behavior would be that the deposited Mn becomes completely detached from the electrode surface due to the evolution of hydrogen. However, the EQCM results show that this explanation may be incorrect.

Through a thermodynamic study using a predominance-existence diagram and Pourbaix diagrams of the Mn species that are liable to exist in solution, it is possible to demonstrate that the reduction of Mn in the pH range 0 to 8.0 occurs via the following reaction:

$$MnCl_3^- + 2e^- \rightarrow Mn^0 + 3Cl^- \quad (11)$$

whose formal potential ($E'_{MnCl_3^-/Mn^0}$) depends on the concentration of Mn(II) and Cl^- ions, as shown in the following equation:

$$E'_{MnCl_3^-/Mn^0} = -1.43 - 0.03pMn' + 0.09pCl' \quad \text{vs SCE} \quad (12)$$

At the working conditions (pMn = 0.222 and pCl = -0.447) the value of the formal potential obtained was:

$$E'_{MnCl_3^-/Mn^0} = -1.48 \text{ V vs SCE}$$

It is clear from this result that, from a thermodynamic viewpoint, the reduction of Mn occurs in the final (most cathodic) part of wave A and with significant interference from the evolution of hydrogen.

Fig. 13 shows the plot of Δm vs E corresponding to the voltammogram (Fig. 12). The offset $\Delta m = 0$ was fixed at the resting potential (E_{rest} = -0.78 V vs SCE).

During the potential scan in the negative direction over the interval of -1.6 to -0.78 V, a mass increase of 2.0 μg cm^{-2} is observed; this increase is much smaller than would be expected from the charge of the voltammogram (Fig. 12) in the same potential interval, an effect that is principally associated with the precipitation of $Mn(OH)_{2(s)}$ onto the electrode surface due to the increase in interfacial pH that occurs during hydrogen evolution, as shown in the following reactions:

$$2H_2O + 2e^- \rightarrow H_{2(g)} + 2OH^- \quad (13)$$

$$MnCl_3^- + 2OH^- \rightarrow Mn(OH)_{2(s)} + 3Cl^- \quad (14)$$

At potentials more negative than -1.6 V, a rapid increase in mass occurs ($\Delta m = 8.22$ μg cm^{-2}). These potentials are sufficiently negative that $MnCl_3^-$ is reduced to metallic Mn. The total cathodic charge obtained from the mass increase (Q^c_{Mn}) in the potential range -1.8 to -0.78 V was 22.82 mC cm^{-2}. From this result, along with the cathodic charge obtained from the voltammogram recorded under the same conditions ($Q^c = 110.67$ mC cm^{-2}), the cathodic efficiency during the reduction of Mn was determined to be 22.73%. This low

efficiency can be attributed to the strong interference of hydrogen evolution in the electrodeposition of Mn.

When the direction of the potential scan was switched to the positive direction starting from -1.8 V, the mass continued to increase up to a maximum value (=19.9 µg cm^{-2}) at E = -1.5 V, after which it remained constant in the interval of -1.5 to -1.33V, forming a plateau. At potentials greater than -1.33 V, a decrease in mass is observed ($\Delta m = -17.82$ µg cm^{-2}) that corresponds to the dissolution of the deposit formed during the scan in the negative direction. At the end of the potential scan, the mass does not return to its original value of zero; instead, 2.23 µg cm^{-2} of mass is accumulated on the electrode surface. It is important to observe that this accumulated mass is similar to the mass recorded during the cathodic scan over the interval of -1.6 to -0.78 V and is associated with the formation of the species $Mn(OH)_{2(s)}$ on the surface of the Fe-EQCM. Thus, it is likely that the $Mn(OH)_{2(s)}$ formed during the potential scan in the negative direction is not dissolved during the inverse scan and remains on the electrode surface.

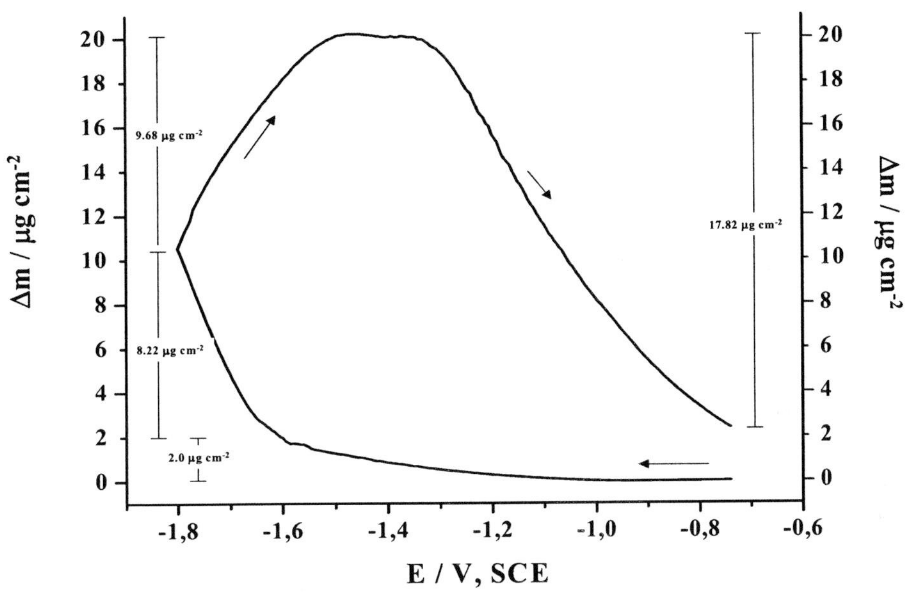

Figure 13. Δm-E curve corresponding to the cyclic voltammogram in Fig. 12.

Comparison of the voltammogram and the plot of Δm vs E reveals that, in the potential interval of -1.5 to -0.78 V during the positive potential scan, there is no oxidation current associated with the loss of mass recorded by EQCM, indicating that the observed dissolution corresponds to a non-Faradaic process.

Massograms provide a graphical representation that affords a clearer picture of the mass changes that occur during potential scans. Figure 14 shows a comparison of the voltammogram and its corresponding massogram for this system. During the potential scan in the negative direction, no significant changes are observed in the mass flux of the massogram in the potential range -1.5 to -0.75V, which verifies that the current observed in the voltammogram in the same potential interval corresponds principally to the hydrogen evolution reaction. At the potential -1.5 V, the positive mass flux (mass gain) begins to increase and continues increasing on going to lower potentials. Given that EQCM does not detect mass changes associated with hydrogen evolution [21], the observed variation in mass is associated exclusively with the deposition of Mn.

On switching the potential scan to the positive direction, two cross-overs are observed between the negative and positive potential scans, behavior characteristic of the formation of new phases [62-63]. The potential at which the cross-over at the less negative potential occurs is known as the cross-over potential (Eco), and occurs when the mass flux is zero. From nucleation theory [63], the cross-over potential in voltammetry can be shown to correspond to the thermodynamic reversible potential of the M^{n+}/M system. In this case the value of Eco was -1.46 V vs SCE. This value is similar to that obtained using the Nernst-type thermodynamic equation (Eq. 12) for the pair $MnCl_3^-/Mn$ ($E'_{MnCl_3^-/Mn}$ = -1.48 V vs SCE), indicating that the feature observed in the massogram corresponds to the reduction of $MnCl_3^-$ to Mn. Thus, the use of massograms provides an alternative method for measuring reversible potentials in the presence of interference from hydrogen evolution.

At potentials more positive than -1.5 V, a negative mass flux (mass loss) due to the dissolution of the deposited Mn is observed. In addition, this dissolution process occurs principally between -1.38 and -0.75 V. It is important to observe that in this potential interval no oxidation current is observed in the voltammogram, indicating that the observed dissolution of the deposit is a chemical reaction [64]:

$$Mn^0 + 2H^+ \rightarrow Mn^{2+} + H_{2(g)} \tag{15}$$

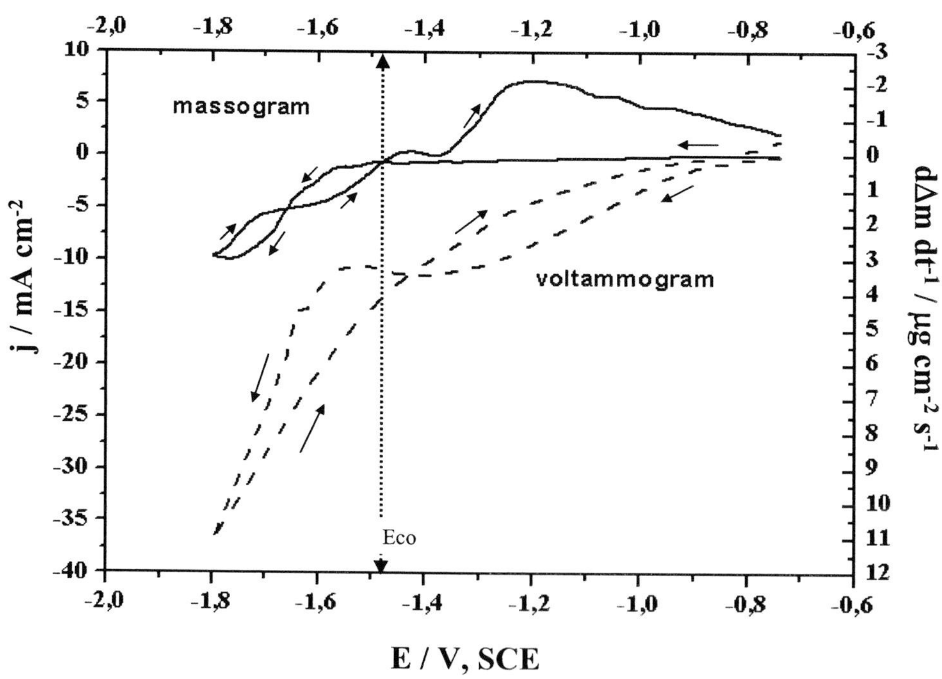

Figure 14. Comparison between the massogram and voltammogram (Fig. 12) obtained for a solution of composition 2.8 M KCl + 0.4 M H_3BO_3 + 0.6 M $MnCl_2$, pH = 5.0.

The above results clearly demonstrate that massograms give information on electrodeposition processes that is difficult to obtain by voltammetry or from plots of Δm vs E. This represents an advantage of massograms over voltammograms for the study of these systems.

5. CONCLUSIONS

The examples described in this chapter demonstrate the power of the EQCM method and the use of massograms in the study of processes occurring at electrode surfaces. In particular, these examples underline the power of the massogram representation proposed by Snook et al. [21], which has important advantages over conventional methods for representing EQCM data. The most important advantage is that it permits direct comparison with the corresponding voltammogram, making it possible to differentiate between mass changes due different types of processes, in particular between changes due to Faradaic and non-Faradaic processes [21]. The results suggest that the routine use of massograms in conjunction with voltammetry will be of great help in studies involving the use of the EQCM.

REFERENCES

[1] Ward, M. D. In *Physical Electrochemistry: Principies, methods and applications*; Rubinstein I. Ed.; Marcel. Dekker, Inc. N. Y., 1995, 293-338.

[2] Sauerbrey, G. *Z. Phys.* 1959, *155*, 206-222.

[3] Nomura, T.; Iijima, M. Anal. Chim. Acta. 1981, *131*, 97-102.

[4] Buttry, D. A.; Ward, M.D. *Chem. Rev.* 1992, *92*, 1355-1379.

[5] Ward, M. In *Physical electrochemistry: Principles, methods and Applications*; Rubinstein I.; Chap. 7; Ed.; Marcel Dekker Inc; N. Y. 1995; 293 - 338.

[6] Hepel, M. In Interfacial Electrochemistry; A. Wieckowski; Chap. 34, Ed.; Marcel Dekker Inc.; N.Y.1999; 599 - 630.

[7] Reed, C. E.; Kanazawa, K. K.; Kaufman, J. H. *J. Appl. Phys.* 1990, *68*, 1993-2001.

[8] Kanazawa, K. K.; Gordon, J. G. *Anal. Chim. Acta* 1985,*175*, 99-105.

[9] Tisionsky, V.; Daikhin L.; Gileadi, E. *J. Electrochem. Soc.* 1996, *143*, 2240-2245.

[10] Schumacher, R.; Borges, G.; Kanazawa, K. K. *Surface Sci.* 1985, *163*, L621-L626

[11] Beck, R.; Pitttermann, U.; Weil, K.G. *J. Electrochem. Soc.* 1992, *139*, 453-461.

[12] Tsionsky, V.; Katz, G.; Gileadi, E.; Daikhin. L. *J. Electroanal. Chem.* 2002, *524*, 110-119

[13] Heusler, K. E.; Grzegorzewski, A.; Jackel, L.; Pietrucha, J. *Ber. Bunsenges Phys. Chem.* 1988, *92*, 1218-1225.

[14] Kelly, J. J.; Anisur Rahman, K. M.; During, C. J.; West., A. C. *J. Electrochem. Soc.* 1998, *145*, 492-497.

[15] Santos, M. C.; Miwa, D. W.; Machado, S. A. S. *Electrochem. Comm.* 2000, *2*, 692-696.

[16] Uchida, H.; Ikeda, N.; Watanabe, M. *J. Electroanal. Chem.* 1997, *424*, 5-12.

[17] Uchida, H.; Miura, M.; Watanabe, M. *J. Electroanal. Chem.* 1995, *386*, 261-265.

[18] Shimazu, K.; Kita, H. *J. Electroanal. Chem.* 1992, *341*, 361-367.

[19] Vatankhah, G.; Lessard, J.; Jerkiewicz, G.; Solfaghari, A.; Conway B. E. *Electrochim. Acta* 2003, *48*, 1613-1622.
[20] Ratieuville, Y.; Viers, P.; Alexandre, J.; Durand G. *Electrochem. Comm.* 2000, *2*, 839-844.
[21] Snook, G. A.; Bond, A.M.; Fletcher S. *J. Electroanal. Chem.* 2002, *526*, 1-9.
[22] Bahena, E.; Méndez, P. F.; Meas, Y.; Ortega, R., Salgado, L., Trejo, G. *Electrochim. Acta* 2004, *49*, 989-997.
[23] Birss, V.; Dang, K.; Wong, J. E.; Wong, R. P. C. *J. Electroanal. Chem.* 2003, *550*, 67-79.
[24] Wu, Q. H.; Sun, S. G.; Xiao, X. Y.; Yang, Y. Y.; Zhou, Z. Y. *Electrochim. Acta* 2000, *45*, 3683-3690.
[25] Xiao, X. Y.; Sun, S. G. *Electrochim. Acta* 2000, *45*, 2897-2902.
[26] Tsionsky, V.; Daikhin, L.; Zilberman, G.; Gileadi, E. *Faraday Discuss.* 1997, *107*, 337-350.
[27] Zilberman, G. *J. Electroanal. Chem.* 2001, *502*, 100-108.
[28] Kouznetsov, D.; Sugier, A.; Ropital F.; Fiaud C. *Electrochim. Acta* 1995, *40*, 1513-1521.
[29] Zilberman, G.; Tsionsky, V.; Gileadi, E. *Can. J. Chem.* 1997, *75*, 1674-1679.
[30] Akiyama, T.; Kobayashi, S.; Ki, J.; Ohgai, T.; Fukushima, H. *J. Appl. Electrochem.* 2000, *30*, 817-822.
[31] Kelly, J. J.; West, A. C. *J. Electrochem. Soc.* 1998, *145*, 3472-3476.
[32] Stoychev, D.; Tsvetanov, C. *J. Appl. Electrochem.* 1996, *26*, 741-749.
[33] Michailova, E.; Vitanova, I.; Stoychev, D.; Milchev A. *Electrochim. Acta* 1993, *38*, 2455-2458.
[34] Trejo, G.; Ruiz, H.; Ortega, R.; Meas, Y. *J. Appl. Electrochem.* 2001, *31*, 685-692.
[35] Yu, J.; Chen, Y.; Yang, H.; Huang, Q. *J. Electrochem. Soc.* 1999, *146*, 1789-1793.
[36] Kim, J.; Lee, J.; Park S. *Langmuir* 2004, *20*, 459-466.
[37] Hope, G. A.; Brown, G. M.; Schweinsberg, D. P.; Shimizu, K.; Kobayashi, K. *J. Appl. Electrochem.* 1995, *25*, 890-894.
[38] Healy,; Pletcher, D.; Goodenough, M. *J. Electroanal. Chem.* 1992, *338*, 155-165.
[39] Bonou, L.; Eyraud, M.; Denoyel, R.; Massiani, Y. *Electrochim. Acta* 2002, *47*, 4139-4148.
[40] Novak, D. M.; Conway, B.E. *J. Chem. Soc. Faraday Trans.* 1981, *77*, 2341-2359.
[41] Zolfaghari, A.; Conway, B.E.; Jerkiewics, G. *Electrochim. Acta* 2002, *47*, 1173-1187.
[42] Birss, V. I.; Chang, M.; Segal, J. *J. Electroanal. Chem.* 1993, *355*, 181-191.
[43] Rand, D.A.J.; Woods, R. *J. Electroanal. Chem.* 1972, *35*, 209-218.
[44] Harrington, D. A. *J. Electroanal. Chem.* 1997, *420*, 101-109.
[45] Watanabe, M.; Uchida H.; Ikeda N. *J. Electroanal. Chem.* 1995, *380*, 255-260.
[46] Uchida H.; Hiei, M.; Watanabe M. *J. Electroanal. Chem.* 1998, *452*, 97-106.
[47] Nicol, M. J.; Philip, H. I. *J. Electroanal. Chem.* 1976, *70*, 233-237.
[48] Aramata, A.; Quaiyyum, Md. A.; Balais, W. A.; Atoguchi T.; Enyo M. *J. Electroanal. Chem.* 1992, *338*, 367-372.
[49] Quaaiyyum, D. A.; Aramata, A.; Moniwa, S.; Sathochi, S.; Enyo, M. *J. Electroanal. Chem.* 1994, *373*, 61-66.
[50] Taguchi, S.; Aramata, A.; Quaaiyyum Md. A.; Enyo M. *J. Electroanal. Chem.* 1994, *374*, 275-277.

[51] Taguchi, S.; Aramata, A. *J. Electroanal. Chem.* 1995, *396*, 131-137.

[52] Horanyi, G.; Aramata, A. *J. Electroanal. Chem.* 1997, *434*, 201-207.

[53] Horanyi, G.; Aramata, A. *J. Electroanal. Chem.* 1997, 437, 259-262.

[54] Aramata, A.; Teuri, S.; Taguchi, S.; Kawaguchi, T.; Shimazu, K. *Electrochim. Acta* 1996, *41*, 761-765.

[55] Mascaro, L. H.; Santos, M. C.; Machado, S. A. S.; Avaca, L. A. *J. Braz. Chem. Soc.* 2002, *13*, 529 -534.

[56] Gonsalves, M.; Pletcher D. *J. Electroanal. Chem.* 1990, *285*, 185-193.

[57] Müller, C.; Sarret, M.; Adreu, T. *J. Electrochem. Soc.* 2002, *149*, C600-C606.

[58] Sagiyama, M.; Urakawa, T.; Adaniya, T.; Hara T.; Fukuda, Y. *Plat. & Surf. Finish.* November 1987, *77*, 77-82.

[59] Govindarajan, G.; Ramakrishnan V.; Ramamurthi, S.; Subramanian, V.; Paryhasaradhy, N. V., *Bull. Electrochem.* 1989, *5*, 422-426

[60] Bozzini, B.; Accardi, V.; Cavallotti, P. L.; Pavan, F. *Met. Finish.* May 1999, *97*, 33-42

[61] Sylla, D.; Creus, J.; Savall, C.; Roggy, O.; Gadouleau, M.; Rafait, Ph. *Thin Solid Films.* 2003, *424*, 171-178

[62] Fletcher, S.; Halliday, C. S.; Gates D.; Westcott, M.; Lwin, T.; Nelson, G. *J. Electroanal. Chem.* 1983, *159,* 267-285.

[63] Fletcher, S. *Electrochim. Acta* 1987, *28*, 917-923.

[64] Gong, J.; Zangari, G. *J. Electrochem. Soc.* 2002, *149*, C209-C217.

In: New Developments in Electrochemistry Research
Editor: Magdalena Nunez, pp. 113-144
ISBN: 1-59454-544-8

Chapter 4

THE DYNAMICS OF FILLING THE FLOW-THROUGH POROUS CATHODE WITH THE METAL TO BE DEPOSITED. MODELING AND EXPERIMENT

A.I. Masliy*, N.P. Poddubny and A.Zh. Medvedev
Institute of Solid State Chemistry and Mechanochemistry of the Siberian Branch of the Russian Academy of Sciences, Michurin str., 15, Novosibirsk, Russia

1. INTRODUCTION

One of the important areas of practical application of liquid-medium flow-through porous electrodes (PE) is metal extraction from diluted solutions [1, 2]. Examples of the problems of this kind and their successful solution due to the use of PE may be the processes of extraction of precious metals from the solutions of hydrometallurgical processing of gold-containing ores and concentrates [3-5], silver recovery from spent photographic solutions [6, 7], metal utilization from washing water and spent electrolytes of electroplating [8].

A distinguishing feature of the performance of PE in metal recovery process is inevitable change in the parameters of the porous matrix with time while it gets filled with the metal. Almost all the parameters of the matrix can change in this process: conductivity, porosity, specific surface. For constant volumetric solution flow rate, these changes will in turn lead to changes in the local linear velocity of liquid movement, and through this parameter and the changed diameter of the fibres of the porous matrix, - to local changes in mass transfer coefficient for the ions of the metal being deposited. The development of the process becomes even more complicated if we take into account the fact that the above-indicated changes occur in every point of the PE at its own rate which depends on the distribution of metal deposition process over the electrode thickness.

* Address for correspondence: Michurin str., 15, 630091, Novosibirsk, Russia. E-mail: masliy@solid.nsc.ru (Masliy A.I.)

From this point of view, the necessity to investigate the dynamics of filling a porous matrix with the metal to be deposited and its effect both on the properties of the matrix itself and on the rate of metal recovery, as well as on its distribution over thickness, is quite obvious. This is important not only for estimating the direction and scale of the influence of separate factors or their set, but also for correct interpretation of experimental results. Nevertheless, it is likely that due to the complicacy of actual account of filling the porous matrix with metal deposit, the major part of theoretical and experimental works on metal electrodeposition on flow-through PE have so far either considered steady models [9, 10] or took into account only the changes of separate parameters with time. For instance, the authors of [11] report equations for the conductivity of the solid phase varying with time, and describe the results of numerical calculations for the case of uniform distribution of current over the PE depth, which is relatively rare. With an increase in the amount of the deposited metal, the conductivity of the porous matrix increases, the increase being most sharp at the initial time of electrolysis. Though this factor is rather important for describing the initial stage of metal electrodeposition on the initially poorly conducting porous matrix, the influence of this factor weakens with time substantially. The effect of other possible variable parameters connected with local changes of filament thickness ($S_v(x)$, $u(x)$, $k_m(x)$) is conserved during the whole electrolysis process and most likely only increases with time. The necessity to take the changes in these parameters during electrolysis into account was stressed in a number of papers [12, 13]; however, we are unaware of any publications with specific description of the dynamic models and analysis of their behaviour. Because of this, in the present work we focus on the account of changes in these parameters with time and over the PE depth; so far, we will not take into account changes in the conductivity of the porous material. When performing specific calculations using the dynamic model, we will restrict the consideration to the analysis of only two limiting cases when metal concentration in solution at the entry into the PE does not change with time (direct-flow mode) and when it, vice versa, decreases substantially (by about an order of magnitude) within the period of PE filling with the metal (circulation regime).

2. The Description of the Dynamic Model

Taking into account the considerations described in the Introduction, and in order to avoid unnecessary complications, we will use the simplest one-dimensional geometric model of the PE. For obviousness, let us represent PE as a set of parallel identical and initially cylindrical fibres (Fig. 1a), along which the solution containing the ions of metal to be deposited flows. In this case, the main parameters of PE are connected with the number of diameter of fibres as follows:

$$\varepsilon = \left(1 - n\pi \frac{d^2}{4 \cdot S_{geom}}\right) \tag{1}$$

$$S_v = \frac{4(1-\varepsilon)}{d} \tag{2}$$

$$u = \frac{m_v}{\varepsilon \cdot S_{geom}} \qquad (3)$$

where S_{geom} is the geometric area of the cross section (accepted to be equal to 1 cm^2), n is the number of fibres, d is their diameter, ε is porosity, S_v is specific surface of the PE, u and m_v are linear and volumetric solution flow rates.

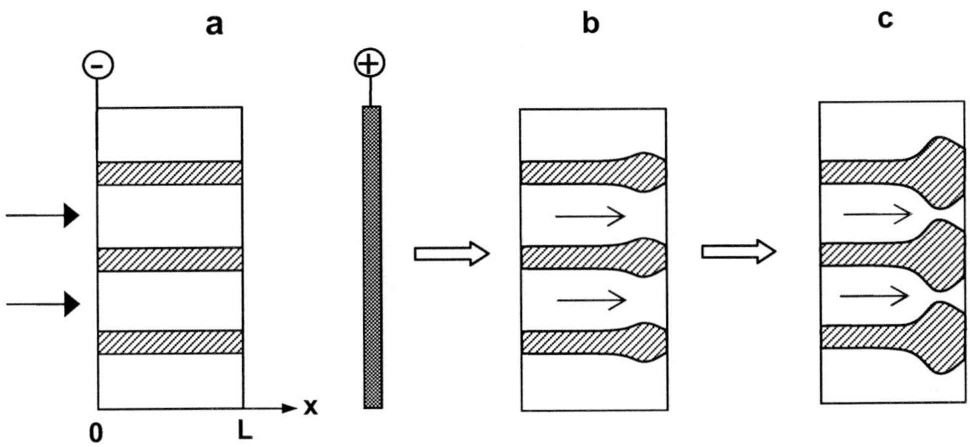

Figure 1. Positional relationship of the porous cathode with rear current feeder and rear solution input, anode and the coordinate axis, and a scheme describing the changes in the shape of filaments of the porous matrix at the initial (a), intermediate (b) and final (c) stages of metal electrodeposition.

In the initial state, these parameters are identical in any point of the PE. However, during the electrolysis, the cylindrical shape of fibres is distorted as a result of non-uniform deposition of the metal, and the fibres turn into more complicated figures of revolution with variable diameter (Fig. 1b). This makes the PE parameters the functions of a point; they can be described by arrays d(x), ε(x), $S_v(x)$ and u(x), which are obtained by applying equations (1)-(3) to the thin cross layers of the porous matrix within the limits of which the indicated parameters may be considered to be constant. Even for constant specific conductivity of solution κ_s and fibre material κ_f, the effective conductivities of the liquid κ_l and solid κ_{sol} phases become variable over the PE thickness. Taking into account the accepted geometry and according to [15]

$$\kappa_l(x) = \kappa_s \cdot \varepsilon(x) \qquad (4)$$

$$\kappa_{sol}(x) = \kappa_f \cdot (1 - \varepsilon(x)) \qquad (5)^1$$

This results in the necessity to use the following system of differential equations for non-uniform porous matrix in order to calculate the distribution of current and potential inside the PE [16]:

---

[1] The account for $\kappa_{sol}(x)$ changes in eq. (5) due to the changes in filament diameter is correct only for the version of the high initial conductivity of the solid phase considered in the present work. For the case of metal deposition on the initially poorly conducting porous matrix, special equations for $\kappa_{sol}(x)$ are to be used.

$$\frac{d}{dx}\left(\kappa_{sol}(x)\frac{dP}{dx}\right) = S_v(x)\cdot i(\eta, c) \tag{6}$$

$$\frac{d}{dx}\left(\kappa_1(x)\frac{d\varphi}{dx}\right) = -S_v(x)\cdot i(\eta, c) \tag{7}$$

with the corresponding boundary conditions:

$$P(0)=0 \tag{8}$$

$$\varphi(0)= \varphi_0>0 \tag{9}$$

$$\left.\frac{dP}{dx}\right|_{x=L} = \left.\frac{d\varphi}{dx}\right|_{x=0} = 0 \tag{10}$$

$$\left.\frac{dP}{dx}\right|_{x=0} = -\frac{i_{ov}}{\kappa_{sol}(0)} \tag{11}$$

$$\left.\frac{d\varphi}{dx}\right|_{x=L} = -\frac{i_{ov}}{\kappa_1(L)} \tag{12}$$

Here P(x) and φ(x) are potentials in the solid and liquid phases; η=P(x)-φ(x) is the potential jump at the interface; c(x) is the concentration of the depositing metal; i(x) is local and i_{ov} is overall density of the total current of the target and side reactions (i(x)<0 and $i_{ov}<0$). The boundary conditions (8) and (9) depict the fact that the electric field potential is determined in each phase to within a constant, and the boundary conditions (10) – (12) correspond to current input into the PE at its front end through the liquid phase and current output through the solid phase at the rear current carrier. It is conditions (12) with the fulfillment of which we determine constant φ_0 in equation (9).

In order to calculate the distribution of current and the concentration of the depositing metal inside the PE, it is necessary to specify the expression for i(η, c), as well as the equation of solution delivery. As we have mentioned in the Introduction, we will consider the case often occurring in practice: the target cathode reaction of metal deposition together with the side reaction of hydrogen evolution proceeding at more negative potential, that is:

$$i(\eta) = i_1(\eta) + i_2(\eta) \tag{13}$$

$$i_1 = i_{01}\left(\exp\left(\frac{\alpha_1 z_1 F}{RT}(\eta - E_1)\right) - \frac{c_s}{c}\exp\left(\frac{(\alpha_1 - 1)z_1 F}{RT}(\eta - E_1)\right)\right) \tag{14}$$

$$i_2 = -i_{02}\cdot\exp\left(\frac{F}{2RT}(E_2 - \eta)\right) \tag{15}$$

Here i_1 and i_2 are current densities of the target and side cathode reactions, i_{01} and i_{02} are exchange currents, E_1 and E_2 are equilibrium potentials of these reactions, α_1 is transfer coefficient, F is Faraday constant, z_1 is the number of electrons necessary for the reduction of the ion of depositing metal, c and c_s are the volume and subsurface concentrations of the ions of the depositing metal in the given section of PE. The kinetic equation for hydrogen evolution takes into account the absence of the concentration limitations, as well as the fact that because of low solubility of the evolved hydrogen for the potential more positive than E_2 $i_2=0$. The transfer coefficient α_2 is accepted to be equal to 0,5.

For the flow-through PE, the system of equations (6)-(15) is supplemented by the equation of the rear delivery of solution:

$$\frac{dc}{dx} = \frac{S_v(x)\cdot i_1(\eta, c)\cdot S_{geom}}{z_1\cdot F\cdot m_v} \tag{16}$$

with the boundary condition:

$$c\big|_{x=0} = c_0 \tag{17}$$

where c_0 is concentration of metal ions at the entrance to the PE. It should be specially noted that equation (16) does not contain porosity $\varepsilon(x)$, because for the model of plug-flow regime a change (for instance, a decrease) in porosity leads to two opposite effects (a decrease in solution volume in a pore, and simultaneous decrease of the time of its residence inside the PE), which compensate each other completely.

In the established state, the concentration term in equation (14) is accounted for through the transfer coefficient of the ions of the depositing metal k_m

$$\frac{c}{c_s} = 1 - \frac{i_1}{i_{1,\,lim}}, \text{ where} \quad (18) \qquad i_{1,\,lim} = z_1 F c(x) k_m(x), \tag{19}$$

$i_{1,lim}$ is the limiting diffusion current with respect to metal ions. When determining k_m, we used the fact that the experimental data on the rate of mass transfer in various porous media depending on the solution flow rate are often approximated by the criterion dependence [17, 18]

$$Sh = a\cdot Re^b, \tag{20}$$

where $Sh = \frac{k_m d}{D_1}$ is Sherwood's criterion, $Re = \frac{u\cdot d}{\nu}$ is Reynolds number, D_1 is the diffusion coefficient of the ions of depositing metal, ν is kinematic viscosity of the solution. Substituting equations for Sh and Re into (20) we obtain the following equation for k_m:

$$k_m = \frac{D_1\cdot a}{d^{1-b}\nu^b}u^b, \tag{21}$$

It follows from this equation that k_m is a power function of flow rate; in the case of fixed composition of the solution (D_l=const, ν=const) it depends also on the filament diameter. The a and b constants were taken from the experimental data for the case of the closest similarity to the case under consideration: a fibrous porous matrix for which the following dependence was found for $D_l=7\cdot10^{-6}$ cm^2/s, ν=0,01 cm^2/s and $d=10^{-3}$ cm [17]:

$$k_m = 0{,}02 \cdot u^{0{,}35} \tag{22}$$

Keeping in mind that the dependencies of (21)-(22) kind are fulfilled for real fibrous materials which are complicated systems of interweaving fibres with their random positions toward the flow direction, including the case of metal electrodeposition [19], one may hope that the character of solution flow around the fibres non-uniformly coated with the metal would not change substantially. So, we used equation (22) to calculate local values of $k_m(x)$ for the corresponding d(x) and u(x) values:

$$k_m(x) = \frac{A}{[d(x)]^{1-b}}[u(x)]^b, \tag{23}$$

where $A = \dfrac{D_l \cdot a}{\nu^b}$,

a=6.4, b=0.35.

The following algorithm was used in numerical calculations of the dynamics of changes in the distribution of current and parameters of the porous matrix. The electrode was divided into (N-1) identical thin layers over its depth (that is, along the length of the filaments shown in Fig. 1); all the parameters of the PE were considered to be identical within each layer[2]. For the initial moment of time, with the d, S_v and u values being the same along the whole length of the fibre, using the solution of the boundary-value problem (6)-(12), (16), (17) we determined the concentrations of the ions of depositing metal c(x), jump of potential at the interface $\eta(x)$, current densities of the side $i_2(x)$ and target $i_1(x)$ reactions in each of N points dividing the filament length. Averaging the values calculated for the edges of segments we obtained the arrays of averaged parameters related to the middles of the layers. These values were considered to be constant during the next small step of time[3].

At each time step dt and for each segment dx, we calculated an increment of metal mass $\Delta m(x)$ deposited inside a given layer by the moment of time t:

$$\Delta m(x) = Q(x) \cdot \frac{M}{z_1 F} \tag{24}$$

and a new filament diameter d(x):

---

[2] The necessary number of division points N was determined using the solution stability criterion for doubling N. For the majority of calculations used in the present work, N=294.

[3] The time of a step was preliminarily chosen so that the final result of the dynamics of filling the porous matrix with metal was independent of step splitting.

$$d(x) = 2\sqrt{\frac{d_0^2}{4} + \Delta m(x)/n\pi\rho dx} \tag{25}$$

Here Q(x) is quantity of electricity consumed for metal deposition in the selected layer:

$$Q(x) = \int_0^t i_1(x) S_{geom} S_v(x) dx\, dt \tag{26}$$

M is molecular mass of the deposited metal, ρ is its density, d_0 is the initial diameter of a filament of the porous matrix.

Further, using the obtained d(x) values, we calculated the new arrays of porosity, linear flow rate, mass transfer coefficient, and specific surface using equations (1), (3) and (23):

$$S_v(x) = n\pi\, d(x)/S_{geom} \tag{27}$$

After that, again the boundary-value problem (6)-(12), (16), (17) was solved, this time with the variable parameters of the porous matrix and deposition conditions, and a new distribution of polarization, current densities and metal concentration over the electrode depth were determined. After that, the cycle of calculating the new set of the PE parameters was repeated. At each time step, the critical point of the electrode was determined in which the thickness of the layer of deposited metal was maximal. The calculation was finished when the total area of surface occupied by the cross sections of the filaments coated with the metal, together with the adjacent thickness of the diffusion layers, approached the geometric area of the PE cross section. For the values of k_m used in our consideration, and for its dependence on filament diameter and flow rate, the residual area of the effective cross-section for the solution flow in the most loaded place of the cathode till the moment of finish of calculation depended on the mean solution flow rate and was about 48 % for u=1 cm/s and even more than 60 % of the geometric area of the PE cross section for u=0,3 cm/s.

In order to check correctness and self-consistency of calculation results at each time step, in addition, it was necessary to fulfill the following conditions with high accuracy (error less than 1%): balance of currents

$$i_{ov} = \int_0^L S_v(x) \cdot i(x) dx \tag{28}$$

and coincidence between two versions of an increase in the mass of the deposited metal, as resulting from equation (16), on the basis of the data on concentration change per pass and on the basis of Faraday's law:

$$m_v(c_0 - c_{out})\Delta t = \frac{I_1 \Delta t}{z_1 F}, \tag{29}$$

where c_{out} is metal concentration in the solution at the outlet from the PE, I_l is total current of metal deposition.

2.1. Estimation of the Relative Contribution of Factors Taken into Account

Before passing to specific analysis of the dynamics of filling with metal for the two limiting cases (direct-flow and circulation modes), let us at first estimate the contribution from the changes in the main factors, taken into account in the model, into the final metal distribution. As we have already indicated describing the model, there are two factors of such an importance: local development of the surface due to filament thickening and local changes in the solution flow rate, as well as changes in mass transfer coefficient of the ions of depositing metal, which is connected with the flow rate and also with the local diameter of the filaments. Correspondingly, Fig. 2 shows the profiles of metal distribution over the depth of PE for identical deposition time and for two L values (0.5 and 1 cm) in three versions of calculation. In the first version (curves 1 and 1′), it was accepted the initial current distribution is conserved during the deposition process, that is, the thickness of deposit in every point increases linearly with time, in agreement with the initial distribution of metal deposition rate. One can see that in this case the smallest amount of metal is deposited and the most gentle slope of its distribution over the loaded zone is observed.

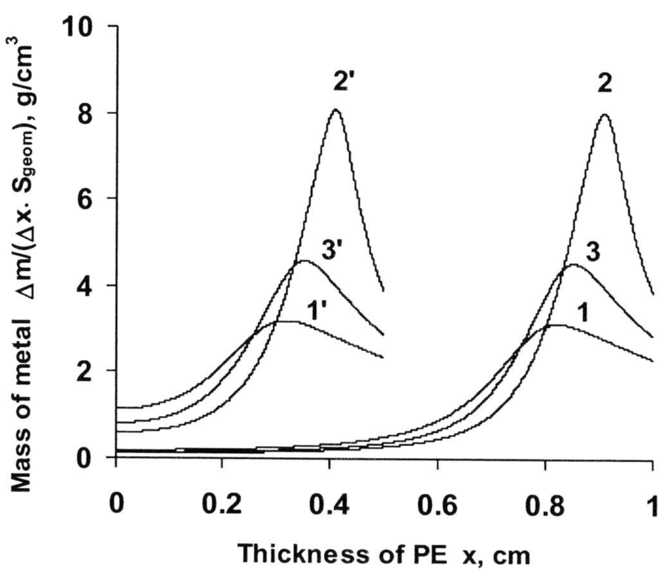

Figure 2. The final profiles of metal distribution over the thickness of the porous electrodes 0.5 cm thick (curves 1O-3O) and 1 cm thick (curves 1-3), calculated for the same duration of electrolysis (2170 s) and three different versions of accounting for the parameters of the porous matrix: 1 and 1O - all the parameters of the distribution of current density are constant and equal to the initial values; 2 and 2O -only changes in the diameter of filament and in porosity (development of surface) are taken into account, but local flow rates and transfer coefficients do not change with time; 3 and 3O - temporal changes of all the parameters of the matrix (S_v, ε and k_m) are taken into account. Conditions for calculation: κ_{sol}=100 Ohm^{-1} cm^{-1}, κ_l=0.2 Ohm^{-1} cm^{-1}, S_v=100 cm^{-1}, m_v=0.975 cm^3/s, c_0=10^{-5} M/cm^3, z_1=1, M=107.87, i_{01}=10^{-3} A/cm^2, α_1=0.5, i_{02}=10^{-6} A/cm^2, E_2- E_1=-0.3 V, i_{ov}=-0.918 A/cm^2, ρ=10.5 g/cm^3.

If one takes into account the dynamics of changes in specific surface $S_v(x)$ but does not consider changes in u(x) and $k_m(x)$, then, as Fig. 2 shows (curves 2 and 2'), this causes a substantial (≈ two-fold) increase in the amount of the deposited metal and its strong localization in the loaded place: rather sharp peak of metal distribution is observed, and its position is noticeably shifted in comparison with curve 1 to the higher polarized front side of the PE. With a decrease in the cathode thickness L from 1 cm to 0.5 cm, qualitative changes in the shape of metal distribution profile and the quantitative correlation of the mass of deposit between the curves 1′ and 2′ are conserved. This example illustrates the importance of taking into account changes in S_v, as well as a substantial contribution from this factor into the dynamics of metal deposition inside the porous matrix and the profile of its distribution over the electrode thickness.

In the case if the changes of all the three parameters ($S_v(x)$, $\varepsilon(x)$ and $k_m(x)$) are taken into account during calculation, then, as one can see in Fig.2, curves 3 and 3′, the deposit profile is obtained which occupies approximately intermediate position between the first two versions, both with respect to the amount of the deposited metal and with respect to the degree of its localization and the position of the maximum. One can also see that, unlike an increase in the filament diameter, this factor acts in the opposite direction and finally slows down metal deposition process substantially and weakens self-localization of the deposit growth process in the loaded zone. So, calculation results indicate that all the factors taken into account in the dynamic model are essentially significant. Because of this, it is important to consider the changes of all the three factors simultaneously, since failure to take into account any one of them leads to a noticeable distortion of the dynamics of filling the PE with metal.

3. Modeling of the Dynamic Process for the Direct-Flow Mode of PE Operation

With the help of the described model, we carried out numerical calculations of the dynamics of filling the PE with metal for different situations. When choosing these situations, we assumed that the essential parameter for the dynamic process under consideration is the degree of non-uniformity of metal distribution at the initial stage of deposition. The latter is most likely determined for high solution flow rate and small reagent transformation degree by the effective thickness of the layer of porous matrix L_{eff}[4] operating in the limiting current mode [20], and its ratio to the total electrode thickness L. Because of this, in the present work we will focus on the results of numerical experiments for u=1 cm/s for three PE thickness

[4] The effectively performing thickness of PE is meant to be a part of its total thickness L proportional to the mean integral value of the ratio of local current of the target reaction $i_1(x)$ to its limiting diffusion value $i_{1,\lim}(x)$ [5]:

$$L_{eff} = \int_0^L \frac{i_1(x)}{i_{1,\lim}(x)} dx$$

In the case of heterogeneous porous matrix formed in the case of non-uniform filling of PE with metal, for the calculation of $i_{1,\lim}(x)$ we take into account not only the changes in metal concentration inside the PE but also the changes in mass transfer coefficient for metal ions due to an increase in the filament diameter and linear solution flow rate.

values (1, 0.5 and 0,3 cm) modeling different ratios of the initial values L_{eff}/L (0.37; 0.72 and 1.0).

The following conditions of PE performance (in addition to earlier mentioned rear current carrier, high flow rate and high conductivity of the solid phase) were conserved the same during calculations:

- the rear reagent feed-in;
- constant concentration of metal ions at the input into the PE;
- kinetic parameters of the target and side cathode reactions (numerical values of the parameters are shown in the legend to Fig. 2);
- overall current density i_{OV}= - 0.92 A/cm^2. [5]

3.1. The Dynamics of Spatial Redistribution of Metal Inside the PE

The profiles of metal distribution calculated after equal time intervals for the limiting L versions (0.3 and 1 cm) corresponding to different initial depth of the penetration of metal deposition process into the PE are shown in Fig. 3 and 4; the values for a number of characteristics of the porous matrix are listed in Table 1.

Table 1. Changes in the characteristics of the porous matrix during electrodeposition (slanting line separates the initial and final values). Here m_0 is the initial mass of the porous matrix, m_f is the final mass of the deposited metal (silver), S_{eff}[6] is the PE surface operating in the limiting current mode.

L, cm	Characteristics of the porous matrix			
	L_{eff}, cm	S_{eff}, cm^2	m_0, mg	m_f, g
0.3	0.30/0.23	30.0/99.1	13.5	1.39
0.5	0.36/0.25	36.2/101.1	22.5	1.55
1.0	0.37/0.26	37.1/101.3	45.0	1.58

[5] It is known that the efficiency of PE performance at first increases with an increase in i_{OV}, and then achieves almost constant limiting value [21]. The moment of transition from the first region to the second one coincides with the start of hydrogen evolution at the most loaded part of the PE. Because of this, the indicated i_{OV} value was chosen to ensure the achievement of limiting current with respect to metal, and also hydrogen evolution, at a part of the electrode, for any version of L, that is, performance conditions close to the limiting ones.

[6] For a homogeneous porous matrix, the effective working surface S_{eff} is directly proportional to L_{eff}: $S_{eff} = L_{eff} \cdot S_v \cdot S_{geom}$, where S_v and S_{geom} are specific and geometric (overall) surfaces of the PE, respectively [22]. In the case of variable specific surface $S_v(x)$, it is necessary to integrate the indicated product over the electrode thickness in order to determine S_{eff}.

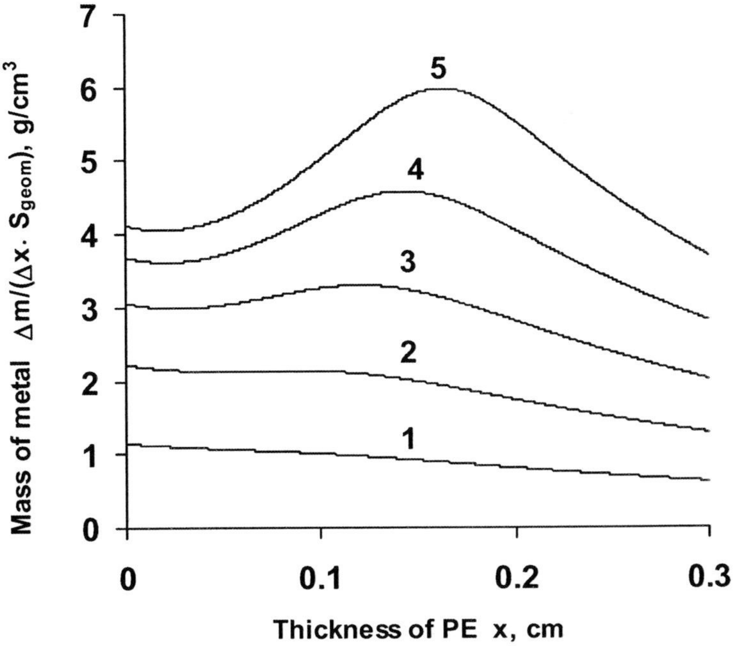

Figure 3. The mass distribution profiles for the deposited metal over the depth of the PE calculated in equal time intervals (s): 1- 423, 2 – 846, 3 – 1269, 4 – 1692, 5 – 2115. The PE thickness L=0.3 cm. Other parameters are the same as those for Fig. 2.

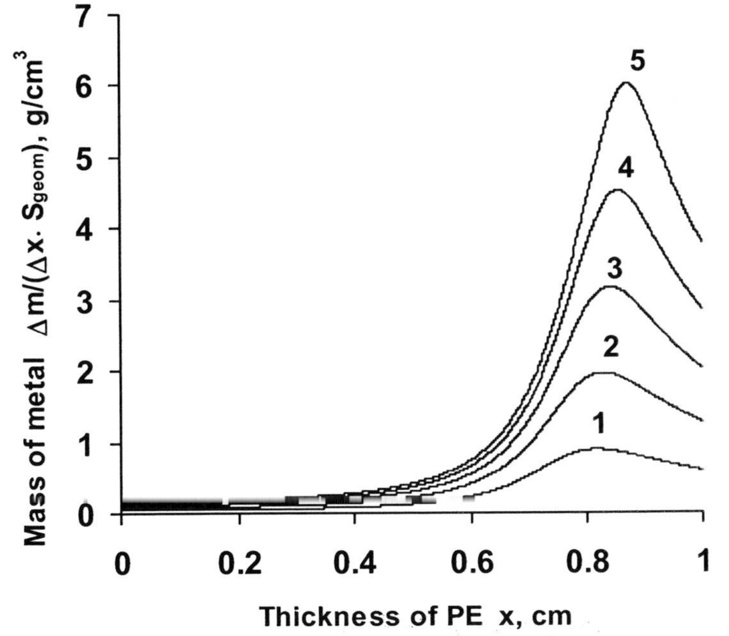

Figure 4. The profiles of metal mass distribution over the depth of the PE calculated in equal time intervals (s): 1- 468, 2 – 936, 3 – 1404, 4 – 1872, 5 – 2340. The PE thickness L=1 cm. Other parameters of calculation are the same as those for Fig. 2.

One can see in Fig. 3 that for $L = L_{eff}$ at the initial stage the metal deposit is distributed over the PE thickness rather uniformly. A small decrease in mass to the front end of the electrode is due to depletion of the solution, so the major part of the metal is deposited near the rear current carrier. However, already curve 3 exhibits a clear maximum in the middle of the PE, increasing with time and shifting to the front end. If we compare an increase in metal mass in different points of the PE (the distance between the neighbouring curves), we can clearly observe a decrease in mass growth near the rear current feeder, while it increases noticeably near the front end and in the middle part of the PE. As a result, the deposition of metal on PE is accompanied by the spatial redistribution and self-localization near the front end which is higher polarizable. Such a trend to redistribution is quite explainable: an increase in the diameter of filaments while they become overgrown with the metal causes a decrease in the effective pore diameter and correspondingly leads to worsening of the uniformity of distribution of the target process and to its localization in the PE region which is the nearest to the anode. Continuing this trend logically till the moment when the neighbouring filaments get closed down and the solution stops flowing through, we should finally obtain a flat electrode with the working front surface instead of a porous electrode.

Qualitatively the same shift of the metal deposition process to the front end of the PE is also observed at larger electrode thickness $L > L_{eff}$ (Fig. 4). Since in this case even at the initial stage the mass of the metal is small at the substantial part of the electrode outside the loaded region ($\approx L_{eff}$) and its increase is only reduced further on, the final distribution of the metal is very non-uniform. Almost the whole metal is concentrated at one fourth of the PE (see Table 1), while the rest space is not used. Nevertheless, a decrease in L_{eff} in the metal deposition on PE does not mean a decrease in its effective working surface. The S_{eff} values at the end of the electrolysis calculated taking the local changes in filament diameter into account are essentially higher (about 3 times) than in the beginning, this growth being almost independent of L. The latter is one more direct evidence that the excessive thickness of the PE layer ($L > L_{eff}$) does not participate in metal deposition process.

It also follows from the data presented in Table 1 that for fixed electrolysis conditions and kinetic parameters the final mass of the deposited metal m_f is almost independent of the PE thickness for $L \geq L_{eff}$. Additional calculations showed that m_f increases with an increase in the difference of the equilibrium potentials of the reactions of metal deposition and hydrogen evolution (that is, the width of the plateau of the limiting current of the target reaction), as well as with a decrease in metal ion concentration and the specific surface. The results of the three series of calculations are represented in Fig.5 in relative coordinates:

$$\frac{m_f}{L} = f\left(\frac{L_{eff}}{L}\right)$$

One can see that all of them fall into one straight line providing an evidence of the proportionality of m_f and L_{eff}.

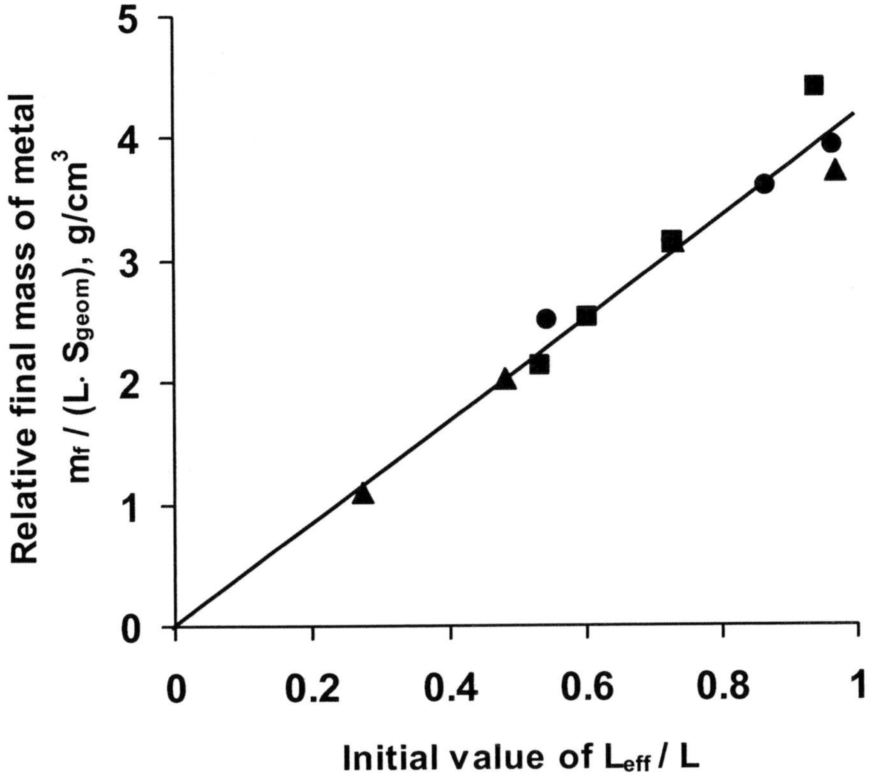

Figure 5. A correlation between the final mass of the metal m_f and the initial L_{eff} values obtained in the series of calculations with variations of the following parameters: ● – E_2, B (-0.1, -0.3, -0.5, -0.7); ■ – S_v, cm^{-1} (50, 100, 150, 200); ▲– $c_0 \times 10^5$, M/cm^3 (0.5, 1, 2, 4). L=0.5 cm. Other parameters are listed in the legend to Fig. 2.

To conclude this section, let us compare the calculated m_f values with the available experimental data. Though absolute m_f values are not too large, with small initial weight of the porous matrix m_0, the ratio of the deposit mass to the matrix mass (which are equal to 103, 68.9 and 35.1 g/g, respectively) even in the case of incomplete filling of the PE cross section with the deposit turns out to be substantially higher than the experimental values (about 50 g/g for gold [23], 20 g/g for silver [24] and copper [25]). Such a deviation can be explained by smaller density of the metals electrodeposited from solutions (especially from diluted ones), and by possible exceeding $L>L_{eff}$ in the experiment, that is, by the excess of unfilled porous matrix.

3.2. The Dynamics of Changes in the Integral Characteristics of PE

Let us now consider the dynamics of changes in PE characteristics averages over thickness. The dependencies of six parameters on electrolysis time are shown in Fig. 6: the mass of the deposit, current efficiency for the metal deposition process, specific surface of the porous matrix, mass transfer coefficient for metal ions, linear flow rate, and the output concentration of the solution for the above-described versions of calculation.

One can see in Fig. 6a that the dependencies of the deposit mass on electrolysis time are approximately linear, though a detailed consideration reveals the S-like shape of these curves: at first, metal deposition rate increases in comparison with the initial one (dotted lines), then passes through a maximum and even decreases slightly at the end of the deposition process. Such a shape of the m=f(t) curve most likely results from the two opposite trends considered above: development of the surface due to an increase in the filament diameter, and a decrease of the working thickness of the PE. The change in metal deposition rate with time is most vividly exhibited for L=0.3 cm and weakens with an increase in the PE thickness. This is connected with the localization of metal deposition process in a limited thin layer, which has been already mentioned, and with smoothing down the effect with an increase of L due to averaging.

Under the chosen electrolysis conditions (constant overall current density), the rate of deposit growth can be changed with time only due to redistribution of the current between the target and side reactions, that is, due to change in current efficiency. Really, as one can see in Fig. 6b, the current efficiency changes non-monotonously during electrolysis: it increases at first (an increase accounts for about 10% for L=0.3 cm and 4-5% for L=0.5 and 1 cm), and slightly decreases at the end. Similarly to the previous case, a decrease in PE thickness due to exclusion of its weakly loaded regions causes a sharper change in the current efficiency. For thicker electrodes, the effect is smoothed down due to averaging.

What is the reason of an increase in the current efficiency of the target reaction as a function of current with the time of electrolysis? It is evident that this is first of all due to the development of the specific surface of the porous matrix due to filament thickening. This causes a decrease in the mean current density and correspondingly results in redistribution of the total current in favor of the target reaction. One can see in Fig. 6c that a substantial increase in the specific surface of the porous matrix occurs in course of metal deposition (even if we take into account the smoothing effect of averaging, an increase is 2-4-fold); the strongest increase is observed at the initial stages of electrolysis. This is due to the fact that within the accepted geometrical model of the PE one and he same amount of the deposit causes a noticeable increase in the filament diameter which is initially small, but has only a weak effect on the mean diameter of the filaments by the end of electrolysis, when the filament diameter is 4 times increased. One of the reasons of a decrease in the current efficiency at the final stage of deposition, along with the mentioned decrease in L_{eff} and replacement of the process to the front end, can also be a noticeable decrease in mass transfer coefficient (Fig. 6e). Comparing the data on the changes in the mean diameter of filaments (Fig. 6c), mean linear flow rate (Fig. 6d) and mass transfer coefficient, one can see that, though the k_m is an increasing function of u and a decreasing function of d, an increase in filament diameter has a stronger effect on k_m.

Non-monotonous time dependence of the current efficiency and the corresponding increase in the mass of the deposited metal leads in agreement with the material balance requirement (see equation (29)) to non-monotonous changes in the output concentration c_{out} (Fig. 6f). At first, when the prevailing effect if the development of the surface, a decrease in the output concentration with time is observed. Al later stages of metal deposition, the negative consequences of filling the porous matrix with metal deposit come into force (a decrease in L_{eff} and k_m), which causes deceleration of the decline and even the subsequent increase in c_{out}.

So, for the case considered in the present section, which involves low degree of metal recovery from solution within one pass through the PE, constant input concentration of the metal in solution and high conductivity of the porous matrix, the proposed dynamic model of PE predicts a number of features, the main of which are:

1) The development of the working surface of PE in the loaded layer, as well as the consequent decrease in the mean current density and increase in the current efficiency at the initial stage of the electrode filling with the metal.
2) A decrease of the depth of penetration of the target process into the PE and of the mass transfer coefficient for metal ions with time, causing localization of the deposition process at the loaded front end and a decrease in metal deposition rate.
3) Non-monotonous time dependence of the current efficiency and metal concentration in solution at the output of PE, depicting the simultaneous action of the above-indicated opposite trends.

4. The Dynamics of Metal Electrodeposition on the Porous Electrode for the Circulation Mode of Its Performance

For the circulation mode of PE performance, a definite volume V of the solution with the initial metal concentration c_0 is continuously circulating between the reservoir and the porous cathode, while the input concentration c_{in} of the metal decreases with time till the required recovery degree is achieved. The scale of metal concentration change in the reservoir during the time of PE filling with metal depends on the ratio of the amount of metal in the reservoir $m_{res}=c_0 \cdot V$ and in the porous matrix filled with the deposit. If $m_{res} >> m_f$, then, within the time necessary for the pores in the critical section of the PE to get filled with the metal and correspondingly for cessation of solution flow, the concentration of the solution in the overhead reservoir and at the input into the porous cathode will be changed insignificantly. Because of this, its change will have only a weak effect on the dynamics of metal redistribution on the PE. It is this simplest version which was analyzed in the previous section.

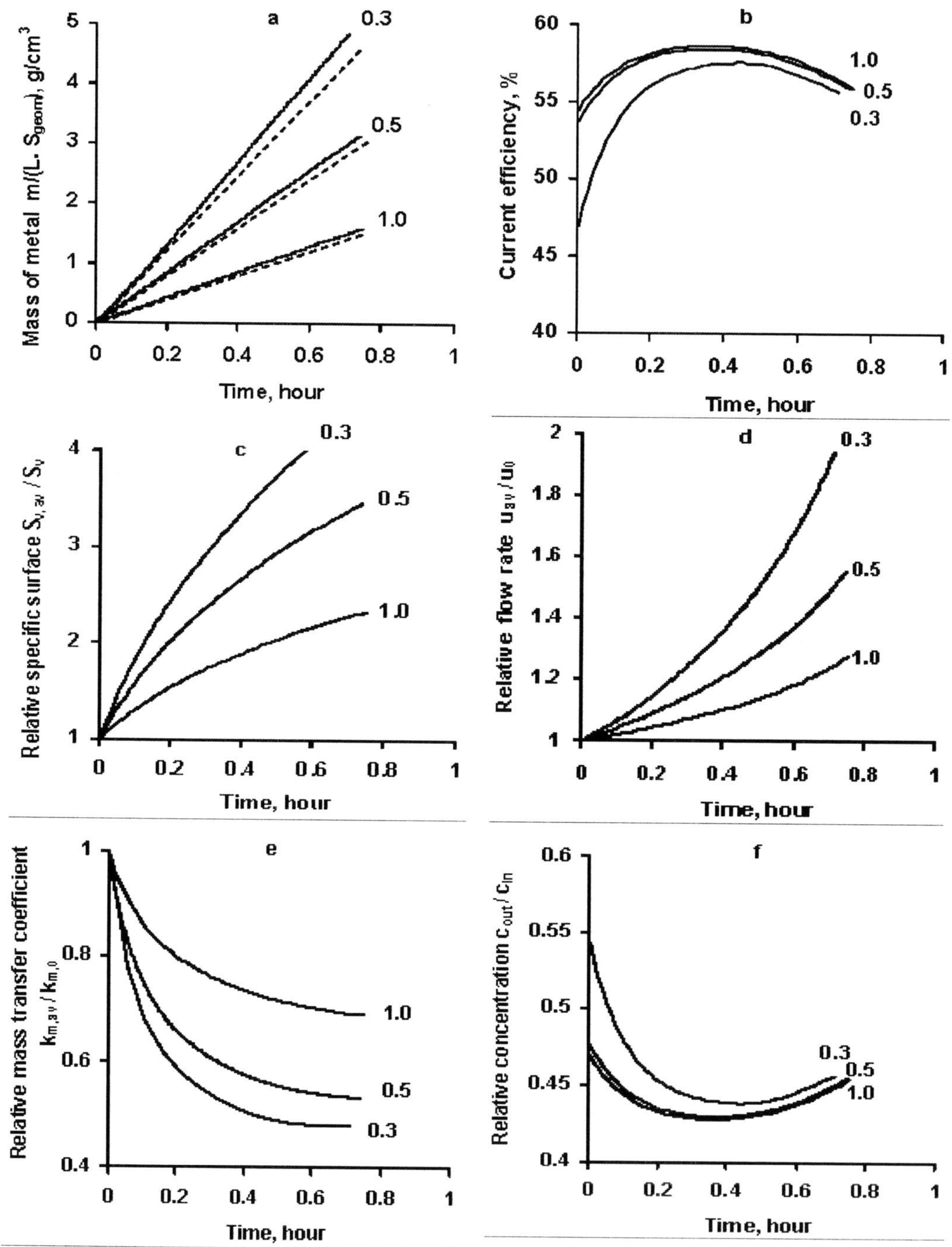

Figure 6. The dynamics of changes in the PE characteristics averaged over L during metal electrodeposition: a) – the deposit mass, b) – the current efficiency of the metal, c) – specific surface, d) – solution flow rate, e) – mass transfer coefficient for the metal ions, f) – the output concentration of metal ions. The figures at the curves correspond to the electrode thickness (cm), other parameters are listed in the legend to Fig.2. Dotted lines in Fig. 6a correspond to the initial slopes of the curves.

If $m_0 \cong m_f$, the input metal concentration will rapidly decrease during filling the porous matrix with the metal. Through changes in the polarization curve of the target process, this decrease should affect the distribution of current and the metal inside the porous cathode, and therefore it should also affect the dynamics of filling it with the deposit. Because of this, investigation of the effect of variable input concentration of the metal in solution is the main problem of the present part of chapter. We will also consider the features of manifestation of this factor for two versions of the process: galvanostatic and potentiostatic. The latter version brings one more variable into the dynamics of PE filling: gradual decrease in the overall current density with time.

For the calculations in circulation mode, the dynamic model of the PE described by us earlier and the calculation algorithm were adapted to this mode by taking into account the changes in the input concentration of the metal during time Δt determined by the balance conditions for metal ions:

$$(c(t) - c(t + \Delta t)) \cdot V = \Delta m, \tag{30}$$

where c(t) and c(t+Δt) are solution concentrations in the reservoir at the moments of time t and (t+Δt), Δm is the mass of metal deposited inside the PE during Δt. We also took into account the natural concentration dependencies of the parameters of polarization curve fort he target reaction: the equilibrium potential and exchange current

$$E_1(t) = E_1(0) + \frac{RT}{z_1 F} \ln[c(t)] \tag{31}$$

$$i_{01}(t) = i_{01}(0)\left[\frac{c(t)}{c_0}\right]^{1-\alpha_1}, \tag{32}$$

where $E_1(0)$ and $i_{01}(0)$ are the parameters at the initial moment of time (while the initial metal concentration is equal to c_0), while $E_1(t)$ and $i_{01}(t)$ are the values at an arbitrary moment of electrolysis (with concentration c(t)corresponding to this moment); α_1 is transfer coefficients; other designations are standard.

In order to simplify the comparison of the results of modeling the circulation and direct-flow modes, we used the same set of PE thickness values (0.3; 0.5 and 1.0 cm) with the same parameters of the porous matrix, as well as the kinetic constants of the target reaction (metal deposition) and side process (hydrogen evolution). The volume flow rate was conserved rather high as in the previous case. 0.975 cm^3/s, as well as the fixed overall current density 0.92 A/cm^2 (for the galvanostatic electrolysis mode).

The volume of circulating solution V was preliminarily chosen so that the metal concentration should decrease by about an order of magnitude till the moment when the calculation is finished.

4.1. Spatial Distribution of the Metal

For a clear view of the dynamics of filling the PE with the metal deposit, 5 profiles of the distribution of the deposit mass calculated for L=0.5 cm in equal time intervals till the finish of calculation are shown in Fig. 7. One can see that, in spite of qualitatively similar shape of the final profile of the deposit distribution, the development of the profile with time in the circulation mode of PE operation has some specific features. First, total increase in the mass of the metal after equal intervals of time (the area between the neighbouring curves) does not increase with time (as in the direct flow mode) but decreases, which is quite natural because of a decrease in c(t). Second, the character of local variations exhibited by an increase in the deposit mass in different points of the PE changes qualitatively. Previously (see Fig. 3 and 4) the growth at the most polarized front end of the electrode and in the point of maximum increased with time; now, in the circulation mode, vice versa, it decreases. On the other hand, a decrease in the metal deposition rate does not occur at the rear end of the electrode, that is, at the solution input site[7], while such a decrease was rather sharp in the case without circulation. As a result, the coordinate of the maximum of the deposit mass shifts not to the front but to the rear end of the PE.

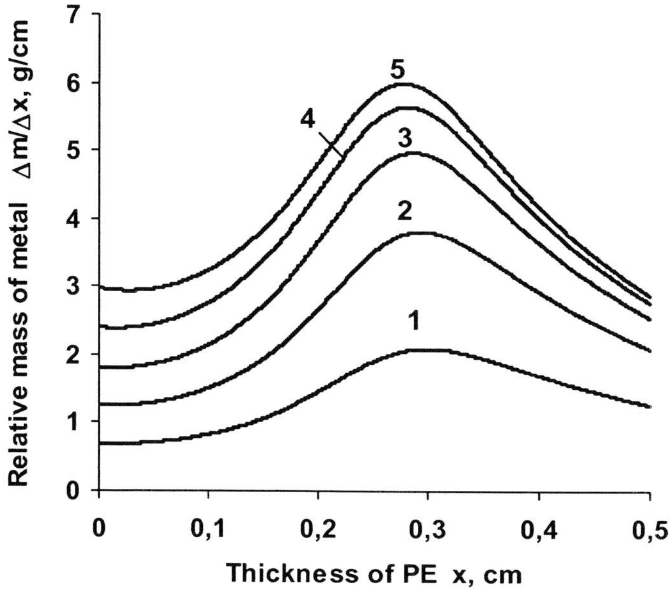

Figure 7. The profiles of the deposit mass distribution inside the PE after equal time intervals (s): 1–1390; 2–2780; 3–4170; 4–5560; 5–6950. Conditions of calculation: circulation mode, V=2260 cm^3, overall PE surface – 1cm^2. The values of other parameters are listed in the legend to Fig. 2.

The final profiles of metal distribution are shown in Fig. 8 for the circulation mode (curves 1-3) and for the direct flow mode (curves 1′-3′) of electrolysis for three chosen L values; numerical values of some characteristics of the electrolysis process are listed in Table 2. It follows from these results that the circulation of electrolyte and permanent decrease in

[7] An increase in metal mass at the rear end versus time passes through a gently sloping maximum within the time interval between curves 3 and 4 in Fig. 7.

the metal concentration in solution have a substantial effect both on the final mass of the metal m_f and on its spatial distribution inside the PE. For $L>L_{eff}$, circulation causes an increase in m_f by a factor of about 1.4, the entire gain being due to the additional deposition of the metal in the rear part of the PE, while, to the contrary, the mass of metal deposited on the most polarized zone near the front end of the electrode decreases. As a result, a characteristic feature of the circulation mode is a shift of the PE section maximally filled with the metal toward the side from which the fresh solution is admitted and, vice versa, weaker metal filling at the most polarized front region of the PE. These features are most clearly exhibited for the thinnest electrode (L=0.3 cm). In this case, the point with the maximal amount of metal shifts till the rear current feeder, while the whole electrode turns out to be in the zone with monotonously decreasing amount of the deposit. As a result, the total mass of the metal for the circulation mode decreases by about 25%.

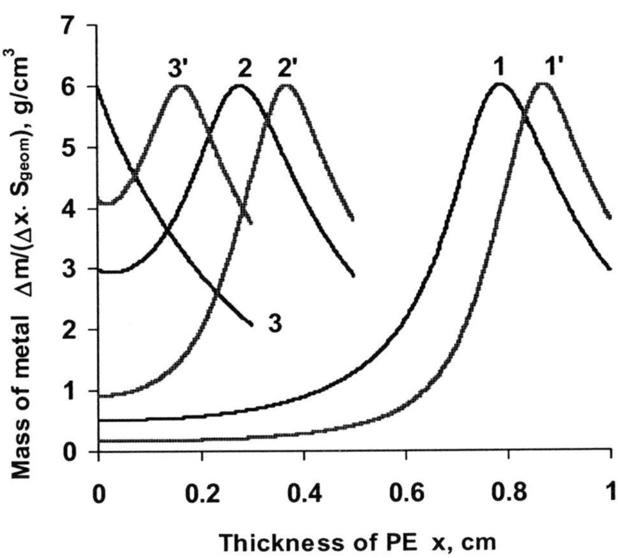

Figure 8. The final profiles of metal distribution for different PE thickness (cm): 1, 1′– 1; 2, 2′– 0.5; 3, 3′– 0.3 for the circulation (1-3) and direct-flow (1′-3′) solution flow modes. The volume of the circulating solution (cm^3): 1– 2330, 2- 2200, 3- 1130. The values of other calculation parameters are listed in the legend to Fig.2.

Table 2. Comparison between the characteristics of metal recovery on PE for the circulation and direct-flow operation modes. Slanting line separates the L_{eff} values in the beginning and at the end of the process

L, cm	Circulation mode		Direct-flow mode	
	m_f, g	L_{eff}, cm	m_f, g	L_{eff}, cm
0.3	1.09	0.30/0.30	1.46	0.30/0.23
0.5	2.11	0.36/0.49	1.57	0,36/0.25
1.0	2.24	0.37/0.76	1.59	0.37/0.26

The reason of the described changes in the spatial distribution of the metal deposit inside the PE is likely to be searched for in the changed conditions of the PE performance. In addition to the factors indicated above for the case of the constant concentration of the metal (first, development of the surface due to thickening of the filaments; second, gradual displacement of the metal deposition process to the front end of the PE), a new factor is added for the circulation mode of solution supply: a substantial decrease in the input concentration of the metal with time. For the case of $L > L_{eff}$, this causes a gradual increase in the effective working thickness L_{eff} and spreading the limiting current mode over the whole electrode. In the case of the rear supply of solution, on the one hand, this provides higher metal deposition rate at the rear part of the PE; on the other hand, as a result of substantial decrease in metal concentration of the solution inside the electrode, this excludes the possibility for the PE to become overgrown (completely filled) with the metal near the rear end. So, under the chosen conditions of circulation (with metal mass in the reservoir being approximately equal to m_f) the new affecting parameter (a decrease in metal concentration of the solution) turns out to be the most important one.

4.2. The Dynamics of Changes in Averaged Characteristics

Though metal deposition process and the corresponding change in the properties of the porous matrix occur at different rates in different points of the PE, as one can see in the previous section, the dynamics of electrode performance as a whole can be characterized by the parameters averaged over the electrode thickness. Calculations showed that the shape of temporal dependencies of the averaged parameters is similar for the electrodes of different thickness. Because of this, the dependencies of six averaged parameters (deposit mass, specific surface, linear flow rate, mass transfer coefficient, metal yield as a function of current and the c_{out}/c_{in} ratio) are shown in Fig. 9a, b only for L=0.5 cm.

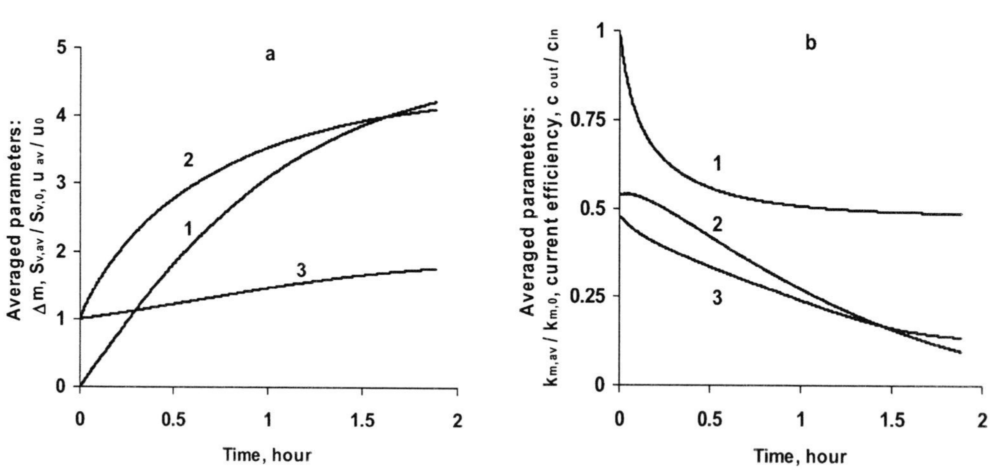

Figure 9. Changes in the PE parameters averaged over thickness during metal deposition for circulation mode:

a) mass of the deposit Δm (g) – 1; relative specific surface $S_{v,av}/S_{v,0}$ – 2; relative linear flow rate u_{av}/u_0 –3.

б) mass transfer coefficient $k_{m,av}/k_{m,0}$ – 1; current efficiency – 2; the ratio of metal concentrations c_{out}/c_{in} – 3.

Comparing these results with the similar data for the direct flow mode (see Fig. 6) one can note the following differences, which are due mainly to a decrease in metal concentration of the solution entering the PE with time:

1) Metal deposition rate decreases with time substantially (almost to zero).
2) As a result, specific surface and linear flow rate increase with time while mass transfer coefficient of the ions of depositing metal decreases, but the decrease is much slower than that without circulation. For instance, identical changes in S_v (by a factor of 4) and u (by a factor of 1.6) were achieved with the direct flow mode within 0.75 hours, while with the circulation mode within 2 hours. At the u_{av}=f(t) dependence, instead of a substantial increase in the growth rate u at the final stage of electrolysis because of a rapid decrease in the distance between the filaments and a decrease in the section free for the solution flow, we observe a decrease in the growth rate u and an approach to some limit, caused by a sharp decrease in metal deposition rate.
3) Unlike non-monotonous temporal dependencies of current efficiency and the c_{out}/c_{in} ratio for the direct flow mode, these parameters decrease monotonously with time in the case of continuous circulation of solution through the PE; the scale of the decrease is much larger. For current efficiency, the main reason of decline is a decrease in the input concentration of the metal in the solution at the constant overall current density, and in the case of c_{out}/c_{in} the most likely reason is an increase in the efficiently working surface of PE due to an increase in polarizability accompanying a decrease in c_{in}.

4.3. The Dependence of Metal Concentration in the Circulating Solution

For the circulation mode of PE operation, the character of changes in the solution concentration in the reservoir versus electrolysis time is determined by the rate of metal deposition inside the PE. If the limiting stage is diffusion of metal ions in the solution and the area of the surface operating under the limiting current conditions does not change with time, then, the change in the solution concentration in the reservoir is described by the following equation [26]:

$$\ln\frac{c_0}{c_t} = \frac{m_v\left[1-\exp\left(\frac{k_m S_{eff}}{m_v}\right)\right]}{V}\cdot t, \qquad (33)$$

that is, it gets linearized in the coordinates ln c_t=f(t) with a slope depending (for constant V and m_v) on the product $k_m \cdot S_{eff}$.

This product does not remain constant during the metal electrodeposition on PE due to some reasons. First of all, a reason is the improvement of polarization distribution inside the electrode with a decrease in metal concentration of the solution, as indicated above [22]. In the initial state, not the whole thickness of the PE operated under the limiting current conditions (L_{eff}<L); L_{eff} increases with time, and so does S_{eff}. Second, the mean diameter of

filaments increases during deposition, which leads to an increase in the working surface area, too. Finally, the third reason of changes in the $k_m \cdot S_{eff}$ product is an increase in the thickness of the diffusion layer when the solution flows around thicker filaments, which should cause a decrease in the mass transfer coefficient k_m with time. Three different versions of calculation are shown in Fig. 10 in order to illustrate the contribution from different factors into the behaviour of the dependence ln $c_t = f(t)$: 1 – according to equation (33) at the initial values of k_m and S_{eff}; 2 – changes in the metal concentration in solution were taken into account, but filament diameter and mass transfer coefficient retained their initial values; 3 – full dynamic calculation taking into account the changes in all the parameters. In addition, the terms of the $k_m \cdot S_{eff}$ product for different moments of time for curve 3 are shown in Table 3.

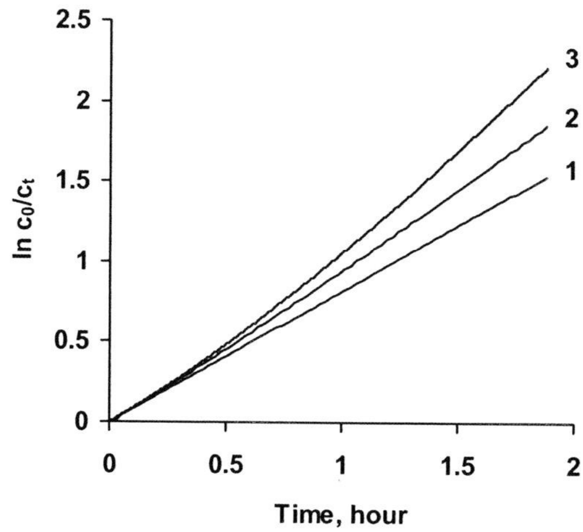

Figure 10. The dependence of metal concentration in the circulating solution on electrolysis time for three versions of calculations: 1 – calculation according to equation (33) for the initial k_m and S_{eff} values; 2 – calculation according to the dynamic model of PE taking into account changes in metal concentration but with constant PE parameters; 3 – full dynamic calculation taking into account the changes in all the parameters. Electrode thickness L=0.5 cm, overall PE surface – 1 cm^2, overall current density – 0.92 A/cm^2, the volume of circulating solution – 2200 cm^3. The values of other parameters are listed in the legend to Fig. 2.

Table 3. The dependence of metal concentration in circulating solution c_t/c_0, PE surface S_{eff} operating in the limiting current mode, and mass transfer coefficient k_m, as well as the product $k_m \cdot S_{eff}$, on electrolysis time. Galvanostatic mode, i_{ov}=0.92 A/cm^2, V=2200 cm^3

t, s	c(t)/c(0)	S_{eff}, cm^2	$k_m \cdot 10^2$, cm/s	$k_m \cdot S_{eff}$, cm^3/s
10	0.998	37.8	1.974	0.746
1350	0.702	82.4	1.176	0.969
2700	0.468	115.7	1.048	1.213
4050	0.299	146.2	1.020	1.491
5400	0.184	180.9	0.928	1.678
6750	0.110	201.0	0.972	1.954

Comparing the curves in Fig. 10 and the data presented in Table 3 one can see that, as a rule, non-linear ln c_t, t dependence with increasing slope is characteristic of the circulation process of metal deposition on the flow-through PE. All the above-indicated reasons contribute into this increase; the effect of the development of surface (an increase by a factor of 5.3) and the effect of a decrease in k_m (approximately 2-fold) partially compensate each other. As a result, total increase in metal deposition rate, as depicted in the changes in the slope of curve 3 Fig. 10, is 2.6 times.

5. The Dynamics of Copper Electrodeposition on Porous Electrode. Experiment

Some recently obtained experimental results on metal distribution inside the PE and on temporal changes in the averaged characteristics of the porous matrix during copper deposition from a diluted sulphuric solution will be reported in this section.

5.1. Experimental Procedure

Carbon felt VINN-250 [27] with fibre diameter 9.5 μm, layer thickness (under the load of 20 g/cm^2) 0.3 cm and specific conductance κ_{sol}=0.1÷0.4 Ohm^{-1}cm^{-1} was used as the cathode porous material. The sample used in the experiment is a round tablet 11.3 mm in diameter with the geometric working surface of 1 cm^2. The initial mass of one tablet is about 30 mg. Since the initial conductance of the porous material is comparable with the conductance of the working solution (κ_l=0.1 Ohm^{-1}cm^{-1}), in order to achieve the condition involved in modeling κ_{sol}>> κ_l we carried out preliminary copper-coating of VINN-250 tablets from sulphuric solution with low copper content (0.1 g/l) and high concentration of the sulfuric acid (100 g/l). Almost complete coating of carbon fibres with copper was observed with a microscope (Neophot-21, Germany) both at the front and rear ends of the electrode. Electric resistance of metallized tablet was about $1\cdot10^{-3}$ Ohm, and the mass of deposited copper was about 10 mg, which accounted for about 2% of the maximal amount of copper which deposited onto one tablet until the moment of fibre accretion in the critical cross section of the PE and cessation of solution flow.

Metallized tablets (one or several; two in the case under consideration) were mounted as the flow-through porous cathode inside the cylindrical holder made of Plexiglas equipped with a current feeder from the rear side (Pt net) and with a stop from the front side (polypropylene net) (Fig. 11). The cathode and anode spaces of the electrochemical cell were separated by an anion-exchange membrane in order to avoid mixing catholyte (0.786 g/l $CuSO_4{\cdot}5H_2O$, 50 g/l H_2SO_4) and anolyte (50 g/l H_2SO_4) and saturation of catholyte with gaseous oxygen. The anode was a disc made of Pt net. The reagents of analytically pure grade and water distilled three times were used to prepare the solutions.

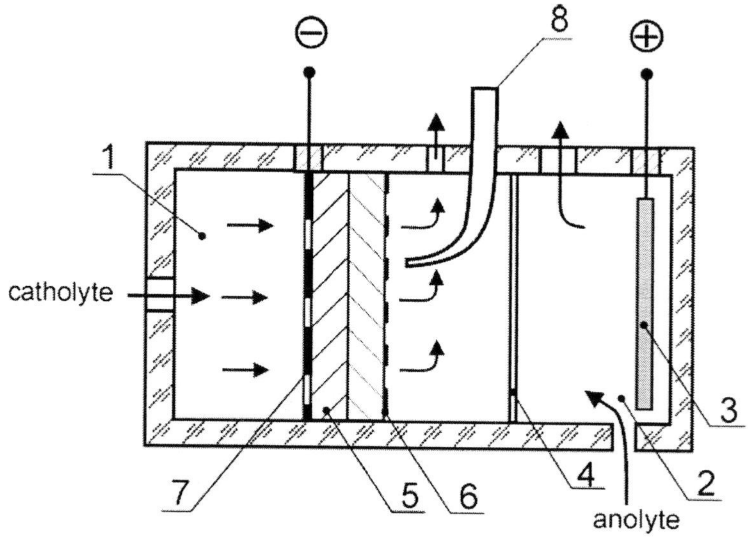

Figure 11. The scheme of the electrochemical cell.

1 – the cathode region of the cell;

2 – the anode region of the cell;

3 – anode (platinum net);

4 – anion exchange membrane of MA-40 type;

5 – porous cathode made of carbon fibre material;

6 – polypropylene net;

7 – current feeder to porous cathode;

8 – titanium net

9 – sealing rubber;

10 – Luggin's capilare to reference electrode

During electrodeposition, the working solution was fed from a reservoir (the solution volume: 5-10 l) with the help of a peristaltic pump into the cathode space of the electrochemical cell and collected in equal portions at the outlet from this space (about 1/20 of the total volume of the solution passed through the PE during the whole experiment); Cu concentration was determined in each portion. The time consumed for collecting each solution sample was also measured; on the one hand, this value was used to calculate the current efficiency of copper as a function of time; on the other hand, it served to monitor the constancy of volume flow rate of the solution. It turned out that the volume flow of solution remained constant until the last portion of the solution (that is, until about 95% of the possible amount of copper was deposited). Only for the deposition of the rest 5% of copper deposit mass, the hydrodynamic resistance of the porous matrix to the solution flow started to increase, which finally made the inlet tubes get torn off the connecting pipes of the peristaltic pump; this was the finish of the experiment. Let us once more stress the difference in the conditions under which the experiment is finished in calculation and in experiment. Calculation was finished when the external boundaries of the diffusion layers of the neighbouring filaments started touching each other and the residual porosity was still about 0.5, while experiment went on until complete filling of the critical section of PE with the metal.

Copper concentration at the outlet of the PE averaged over the sampling time was determined by means of trilonometric titration [28]. Knowing the difference in the concentrations at the inlet and outlet of the PE, and the sample solution volume (V_{sam}), we determined an increase in copper mass for a definite interval of time $\Delta m = \Delta c \cdot V_{sam}$; taking into account Δm and the consumed quantity of electricity we calculated the current efficiency of copper as a function of time.

In order to measure an increase in mass and metal redistribution profile over the depth of the PE, a PE composed of several parts is usually used [29,30]. Electrolysis process is ceased from time to time, the PE is separated into the composite parts, washed, dried and weighed. This is a complicated procedure, which also can bring about some changes into the dynamics of metal deposition. In addition, it tuned out that the separation of layers of a composite PE can be easily carried out only in the case of small amount of the deposited metal. At later stages of deposition, the neighbouring layers of the PE grow so closely together that it is impossible to separate them from each other. The boundary between the layers can hardly be distinguished at the cross section. Because of this, in the present work we measured mean diameter of a fibre in the sections parallel to the current feeder and located at different distances from it in order to study the distribution of copper deposit inside the PE.

For this purpose, after cessation of copper deposition, thorough washing and drying, the porous cathode was impregnated with epoxy resin under pressure (in order to avoid the formation of air bubbles). After polymerization of the resin, a cylindrical sample was cut along its diameter; after standard preparation of the surface of cross section, it was placed on the stage of microscope so that the plane of the rear current feeder coincided with the longitudinal axis of the ocular crossing. Moving the sample at a definite distance from the current feeder line with the help of the tuning micro-screw we determined the diameters of all the filaments crossed by the longitudinal axis of the ocular. A characteristic view of changes in the cross section of the PE as the distance from the rear current feeder increases is shown in Fig. 12. Having thus determined the mean outer diameter of the filaments in a given section and taking into account the initial diameter of the carbon filament, we calculated the mean area of the metal ring which was considered to be proportional to the mass of metal deposit.

5.2. Experimental Results and Discussion

In the present section we will describe experimental results obtained for galvanostatic copper deposition (i_{ov}=0.35 A/cm^2) in the direct-flow mode (input copper concentration: 0.2 g/l) at the fixed volume flow rate of 0.8 cm^3/s with a PE composed of two layers of preliminarily metal-coated carbon felt VINN-250 having the geometric surface of 1 cm^2.

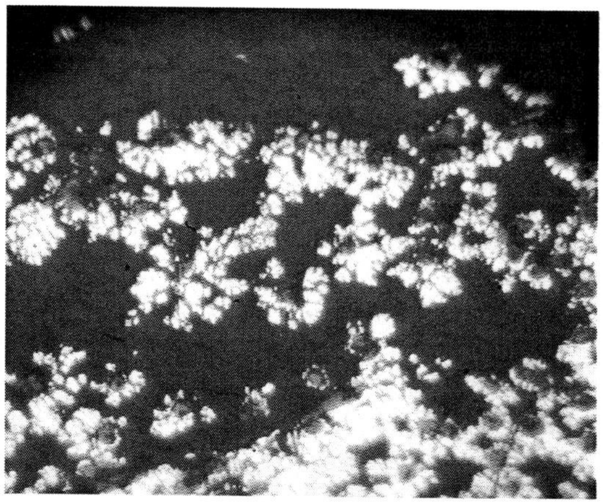
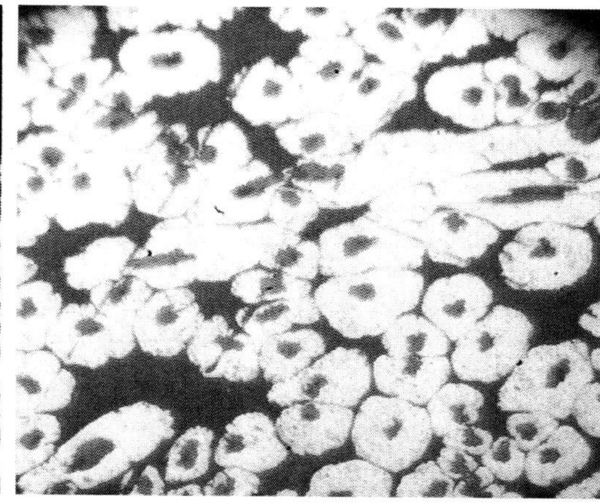

a b

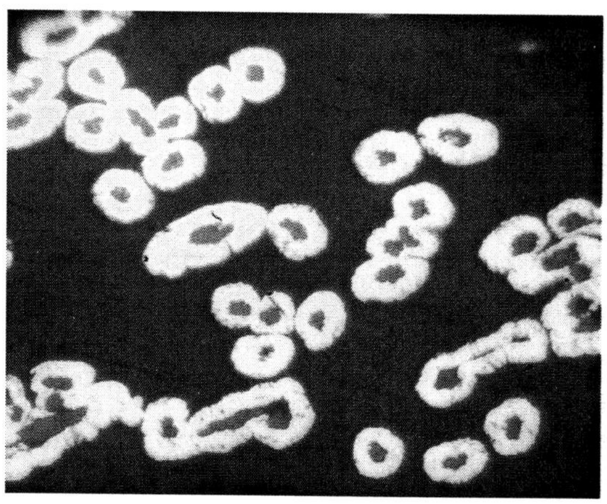
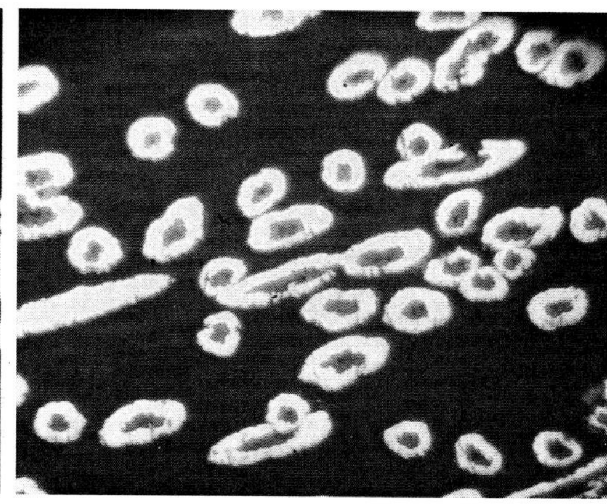

c d

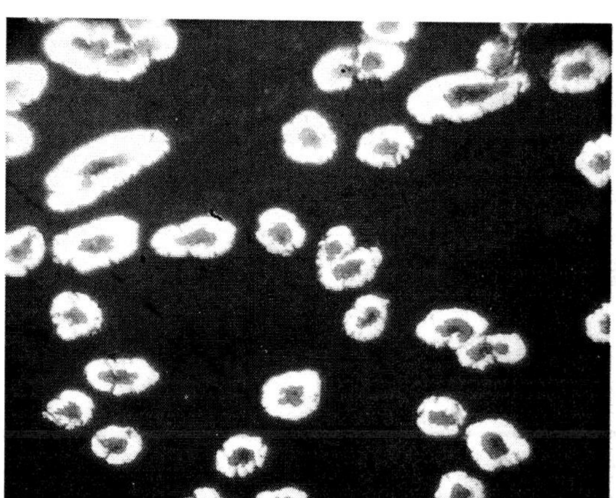
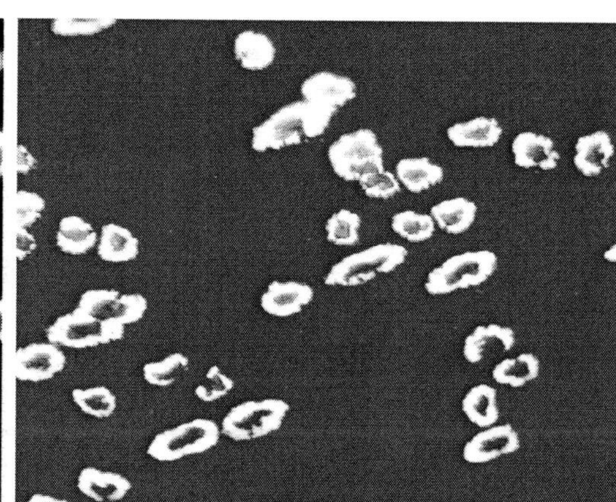

e f

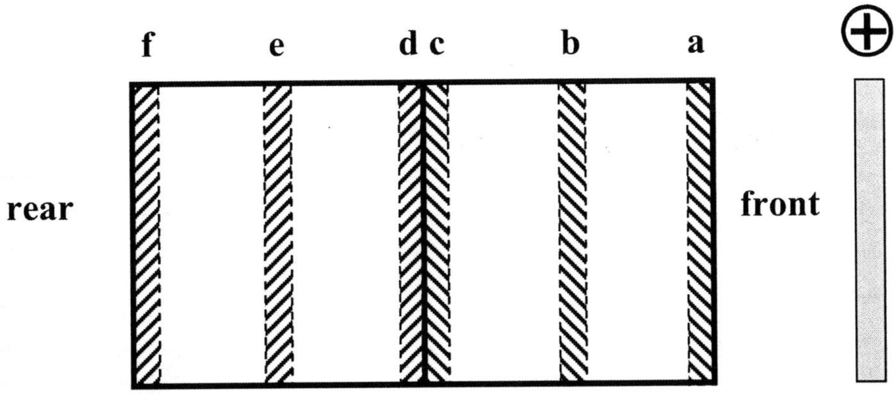

Figure 12. Photomicrographs of the regions of cross section of copper-filled PE made of two VINN-250 layers; the regions are situated at different distances from the rear current feeder. The position of the longitudinal line of the ocular for each image is indicated in the scheme.

The final distribution of the metal determined using the above-described procedure is shown in Fig. 13. It follows from the obtained profile that the major part of copper deposited in higher polarized PE layer which is close to the anode. However, inside this layer, too, metal distribution is very non-uniform. The section filled with the metal to the highest extent is situated near the front end of the PE. Both these facts are in good qualitative agreement with the results of modeling for the case when $L \geq L_{eff}$ (for example, see Fig. 4) and obviously demonstrate the localization of the critical section at the most polarized end of the PE, as well as insignificant effect from the additional rear layer.

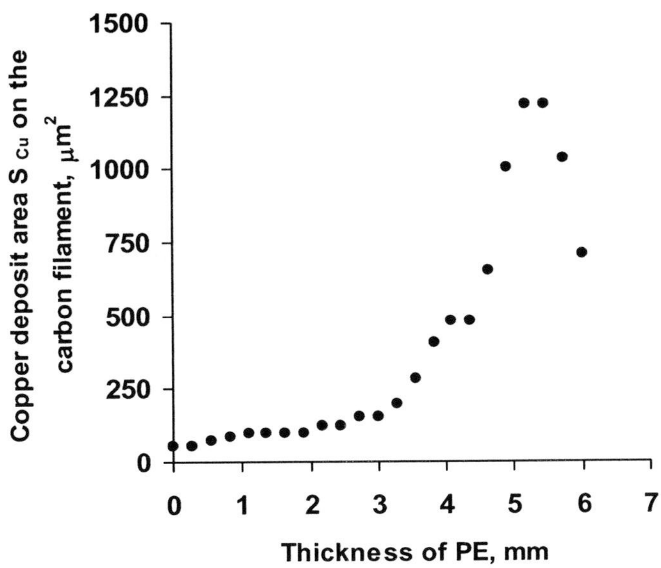

Figure 13. Changes of the effective copper deposit area S_{Cu} on the carbon filament over the depth L of PE made of two VINN-250 layers.

Discussing the profile of the final metal distribution one should pay attention to the photomicrography of the cross section of the PE region which is the nearest to the anode (Fig. 12a). This region is a very friable porous copper deposit in which one can hardly find carbon fibres and estimate their diameter. This layer most probably depicts copper deposition at the final stage of electrolysis onto the front part of the PE under the worsening solution flow conditions and high local current density.

The experimental dependencies of copper concentration at the outlet of the PE, current efficiency and mass of the copper deposit on deposition time are shown in Fig. 14 (a,b,c). Comparing these curves with the corresponding calculated dependencies of the direct-flow mode (see Fig. 6) one may conclude that the shape of experimental curves is similar to that of the calculated ones at the initial stage of electrolysis. At first, copper concentration at the outlet of the PE slightly decreases (while current efficiency and copper overweight correspondingly increase), passes through rather shallow minimum, and then, after approximately 1 hour, c_{out} starts to increase (while current efficiency and copper overweight start to decrease). So, the character of changes in the averaged parameters of the PE at this stage of copper deposition confirms qualitatively one of the important conclusions obtained by modeling (for the above-indicated specific conditions of experiment and calculation), that is, their non-monotonous change. This non-monotonous character depicts the effect of two oppositely acting processes: development of surface due to metal growth over the filaments and gradual displacement of the deposition process to the front end of the PE because of a decrease in pore diameter. In our opinion, this is rather important fact providing evidence that, in spite of many substantial simplifications in describing the PE geometry, hydrodynamics of solution flow and total polarization curve (failure to take into account solution filling with gas due to hydrogen evolution), the proposed dynamic model provides qualitatively correct description of the dynamics of redistribution of current, potential and metal concentration inside the PE.

The experimental dependencies of c_{out}, current efficiency and metal mass at the second half of the electrolysis process on time are of special interest because they characterize the phenomena occurring in the PE at small (till zero) porosity inaccessible for calculation. One can see in Fig. 14 that two clearly different regions are observed in this deposition time region on the c_{out}=f(t) and CE=f(t) curves. A very noticeable increase in c_{out} and, vice versa, decrease in current efficiency and the rate of deposit mass increase occur at the first region. This means a more clearly exhibited trend to a decrease in the working surface due to a decrease in the depth of penetration of the copper electrodeposition process inside the PE, which was revealed as far back as at the previous stage of electrodeposition; this coincides with the predictions of the model.

At the final stage of electrolysis, the process is stabilized; the parameters under discussion change only weakly with time, which is an evidence of approximate constancy of the working surface of PE during this period. Taking into account the features of metal distribution near the front end of PE (Fig. 12a), one can assume that the final region of curves in Fig. 14 depicts approximate equality between the continuing decrease in the working surface and its simultaneous development due to the growth of a friable and rough deposit directly at the front end of the electrode. It should be noted that this effect was not foreseen in the proposed model: current input from the solution directly into the front end of the PE is prohibited. Current may go into the solid phase only through the side surface of the filaments.

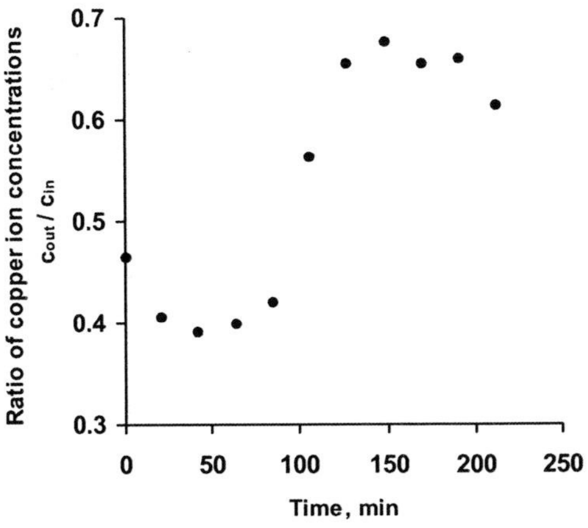

(a)

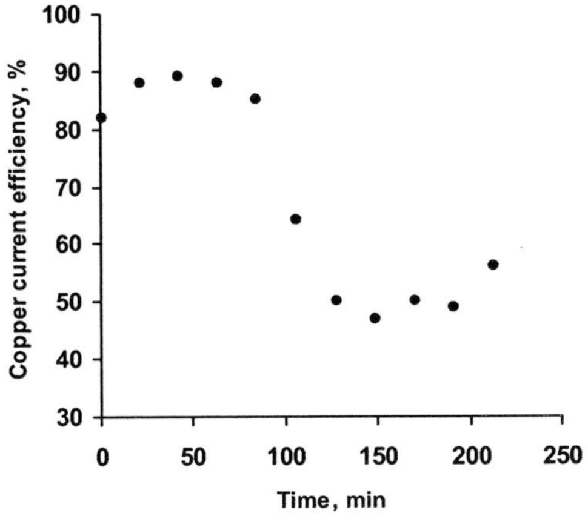

(b)

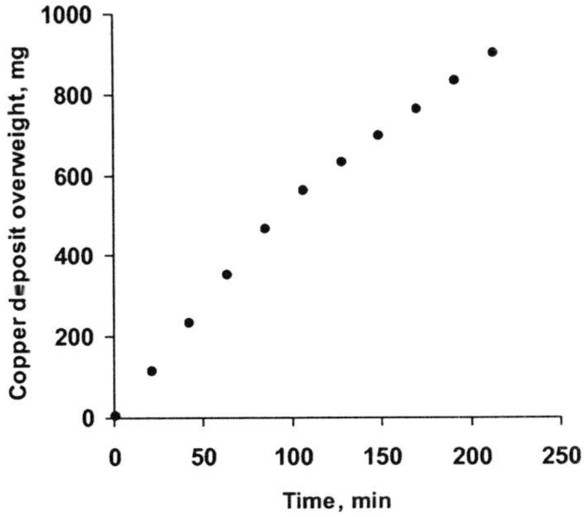

(c)

Figure 14. Changes during electrolysis: a– the ratio of copper ion concentrations in solution at the outlet and inlet (c_{in}=0.2 g/l =const) in the PE; b– copper current efficiency; c– copper deposit overweight.

ACKNOWLEDGEMENTS

The authors thank Prof. R.Yu. Beck for useful remarks in discussing the results, M.A. Kostirya for assistance in carrying out calculations and in paper preparation, and E.V. Sukhanova for assistance in carrying out experiments.

The work has been supported by RFBR (Project No. 04-03-32150).

REFERENCES

[1] Bek R.Yu. //*Izvestiya SO AN SSSR. Ser. khim. nauk.* 1977. No. 4. Issue 6. P.11 (in Russian).

[2] Newman J.S. and Tiedemann W. //*Adv. Electrochem. and Electrochemical Eng.* 1978. V.11. P.353.

[3] Masliy A.I., Bek R.Yu., Makhnyr N.V., Ganin V.M., Shashin V.A. //*Tsvetnye metally (Nonferrous metals).* 1973. №8. P.73 (in Russian).

[4] Masliy A.I., Zamyatin A.P., Makhnyr N.V. //*Izvestiya SO AN SSSR. Ser. khim. nauk.* 1979. №4. Issue 1. P.113 (in Russian).

[5] Varentsov V.K. // *Izvestiya SO AN SSSR. Ser. khim. nauk.* 1984. №17. Vip.6. P.106 (in Russian).

[6] Varentsov V.K., Varentsova V.I. //*Chemistry for Sustainable Development.* 1996. V.4. №2. P.181.

[7] Masliy A.I., Vais A.A., Panasenko A.A., Kuzmenko L.A. //*Chemistry for Sustainable Development.* 1997. V.5. №5. P.339 (in Russian).

[8] Varentsov V.K. //*Zhurnal ekologich. khimii.* 1993. №4. P.335 (in Russian).

[9] Van Zee J.//*J.Electrochem. Soc.* 1977. V.124. P.706.

[10] Trainham J.A. and Newman J.S. //*J. Electrochem. Soc.* 1977. V.124. №10. P.1528.

[11] Beck R.Yu., Zamyatin A.P., Koshev A.N., Poddubny N.P. // *Izvestiya SO AN SSSR. Ser. khim. nauk.* 1980. №2. Issue 1. P.110 (in Russian).

[12] Koshev A.N., Varentsov V.K., Gleizer G.N. //*Elektrokhimia.* 1992. V.28. №8. P.1128 (in Russian).

[13] Koshev A.N., Davydenko A.A., Varentsov V.K., Gamburg V.G., Gazeeva N.V., Troyan G.F. // *Elektrokhimia.* 1997. V.33. №1. P.20 (in Russian).

[14] Vedenjapin V., Baturova M., Danilova E., Weichgrebe D. //*Abstracts of 202-nd Electrochem. Soc. Meeting.* 20-24 Oct. 2002. Salt Lake City. Abstr. №241.

[15] Masliy A.I., Poddubny N.P. //*J. Appl. Electrochem.* 1997. V.27. №9. P.1045.

[16] Daniel-Bek V.S. //*Zh.F.Kh.* 1948. V.22. №6. P.697 (in Russian).

[17] Beck R.Yu. //*Sib. Khim. Zhurn.* 1993. №3. P.85 (in Russian).

[18] Shamal D., Van Erkel J., Van Duin P.J. //*J. Appl. Electrochem.* 1986. V.16. №3. P.422.

[19] Beck R.Yu., Zamyatin A.P. //*Elektrokhimia.* 1978. V.14. №8. P.1196 (in Russian).

[20] Masliy A.I., Poddubny N.P. //*Elektrokhimia.* 1994. V.30. №7. P.897 (in Russian).

[21] Masliy A.I., Poddubny N.P. //*Elektrokhimia.* 1995. V.31. №12. P.1398 (in Russian).

[22] Masliy A.I., Poddubny N.P., Panasenko A.V. //*Elektrokhimia.* 1997. V.33. №1. P.101 (in Russian).

[23] Masliy A.I., Zamyatin A.P., Varentsov V.K., Krapivin V.M., Frolov Yu.I. //*Tsvetnye metally* (Non-ferrous metals). 1976. №8. P.34 (in Russian).

[24] Varentsov V.K. // in: «*Intensification of electrochemical processes*» ed. by A.P. Tomilov. Moscow: Nauka, 1988. P. 94-118 (in Russian).

[25] Zherebilov A.F., Lukyanov V.O., Varentsov V.K. //*Tsvetnye metally (Non-ferrous metals)*. 1983. №4. P.31 (in Russian).

[26] Beck R.Yu., Masliy A.I., Lavrova T.A. // *Izvestiya SO AN SSSR. Ser. khim. nauk*. 1972. Issue 1. P. 25 (in Russian).

[27] Varentsov V.K., Zherebilov A.F., Maley M.D. // *Izvestiya SO AN SSSR. Ser. khim. nauk*. 1984. №17. Issue 6. P. 120 (in Russian).

[28] Charlo G. *Methods of Analytical Chemistry. Moscow. Khimia*, 1965 (in Russian).

[29] Bek R.Yu., Zamyatin A.P. //*Elektrokhimia*. 1980. V.16. №11. P.1316 (in Russian).

[30] Zherebilov A.F. Regularities of metal electrodeposition on flow-through cathodes made of carbon fibre materials. *Thesis Cand. Sci. Novosibirsk*. 1987 (in Russian).

In: New Developments in Electrochemistry Research
Editor: Magdalena Nunez, pp. 143-183

ISBN: 1-59454-544-8

Chapter 5

SOLID OXIDE ION CONDUCTORS: MATERIALS AND APPLICATIONS

V. Thangadurai[a,b*] ***and W. Weppner***[b]

[a,b]Department of Chemistry,University of Calgary,
2500 University Dr. N. W., Calgary, AB, T2N 1N4, Canada
Phone 001 403 210 8649: Fax. 001 403 289 9488
[b]Chair for Sensors and Solid State Ionics, Faculty of Engineering
University of Kiel, Kaisersstr. 2., D24143 Kiel, Germany
Phone: 0049 431 880 6210, Fax. 0049 431 880 6203

ABSTRACT

The state-of–the-art of the current solid oxide fuel cell (SOFC) materials development is presented and major problems with the present materials are briefly discussed.

Perovskite-type (ABO_3) structured metal oxides have been explored to find appropriate materials for intermediate temperature SOFCs (IT-SOFCs) and other ionic devices that include chemical sensors for oxygen and hydrocarbon. Replacement of divalent Sr^{2+} ions by monovalent K^+ ions in $La_{0.9}Sr_{0.1}Ga_{0.8}Mg_{0.2}O_{2.85}$ (LSGM) decreases the electrical conductivity. Substitution of trivalent La^{3+} by substantial amounts (~50 atom-%) of trivalent Pr^{3+} in LSGM and the In-analogue of LSGM, $La_{0.9}Sr_{0.1}In_{0.8}Mg_{0.2}O_{2.85}$ (LSIM), yield new mixed oxide ion and electronic conductors. Among the investigated compounds, $La_{0.4}Pr_{0.4}Sr_{0.2}In_{0.8}Mg_{0.2}O_{2.8}$ exhibits the highest total electrical conductivity of ~ 3.2×10^{-3} S/cm at 200 °C in air with the lowest activation energy of 0.44 eV (200-700 °C). Oxygen partial pressure dependence of the electrical conductivity measurements show that LSIM and Pr-substituted LSGM and LSIM perovskites are p-type electronic conductors at high oxygen partial pressures (10^{-5} – 0.21 atm) at 750 °C.

Solid solutions $SrSn_{1-x}Fe_xO_3$ exhibit compositionally driven phase transitions. For example, the composition x = 0 to 0.3 crystallizes in an orthorhombic structure, x = 0.4–0.9 crystallizes in simple cubic and x = 1 in orthorhombic structure. The electrical

* For correspondence E-mail. vthangad@ucalgary.ca

conductivity of $SrSn_{1-x}Fe_xO_3$ perovskites increases with the increase in the Fe content and reaches a value of about 10^{-1} S/cm at 25 °C in air for x = 1. Compounds with low Fe content exhibit both ionic (O^{2-}) and electronic conductivity, while the higher Fe content (≥ 0.5) perovskites are predominate electronic conductors over a wide range of oxygen partial pressures (10^{-22}–0.21 atm) at 750 °C.

Hebb-Wagner (HW) measurements in the oxygen partial pressure range 10^{-3}- 1 atm using an oxide ion blocking electrode show that $Ce_{0.8}Sm_{0.2}O_{1.9}$ (CSO) is predominantly oxide ion conducting in the temperature range 500-600 °C and its excess electronic (hole) conductivity is about three orders of magnitude lower than the bulk ionic conductivity. The electronic (hole) conductivity is proportional to $P_{O2}^{1/4}$. The diffusion coefficient of the electrons is found to be about 10^{-6} cm^2/sec at 500-600 °C. This value is of the same order of magnitude as the diffusivity (component diffusion coefficient) of oxygen reported for Gd-doped CeO_2 (CGO) (~10^{-6}-10^{-5} cm^2/sec at 500-970 °C). The activation energy for the electronic conductivity is found to be 1.2 -1.6 eV in the oxygen partial range 10^{-3}-1 atm, which is comparable to those of other rare earth (RE)-doped CeO_2. Application of CSO in amperometric oxygen sensors at 550 °C has been demonstrated. The limiting current varies according to Fick's law in the oxygen concentration range between 100 and 500 ppm.

Scheelite-type ABO_4 (A = Ca, Sr, Ba, Pb; B = Mo, W) powders of the shape of small spheres (except $BaBO_4$: needles) were prepared at room temperature by metathetic reactions using aqueous solutions of alkaline or lead chloride, and sodium molybdates or tungstates. The open circuit voltage (OCV) measurements reveal ionic transport with an average transference number of 0.86-0.95 over the oxygen partial pressure range 10^{-4} - 10^{-19} atm in the temperature range 600 – 850 °C for $SrWO_4$.

1. Introduction

Ionic conduction in solids has been a well-known phenomenon since the 1890s. However, fast progress in understanding and developing new materials has been achieved only in recent years. Much attention was paid to solid electrolytes, which are ionic conductors with commonly a single predominantly conducting species (anionic or cationic) and have a negligibly small electronic (electrons and holes) conduction. Several solid electrolytes (metal oxides, non-oxides, composites and organic polymers) are known for proton, lithium, sodium, potassium, silver, copper, fluoride, and oxide ion conduction. Solid electrolytes exhibit conductivity values in the range 10^{-1}-10^{-8} S/cm at room temperature, which is comparable to that of semiconductors and aqueous electrolytes (Fig.1) [1-8].

Recent research on solid electrolytes has drawn much attention due to their potential possible technological applications in various solid-state ionic devices that include solid oxide fuel cells (SOFCs), proton membrane fuel cells (PMFCs), sensors for various chemical species (e.g., H_2, O_2, CO, CO_2, NH_3, hydrocarbon), electrochromics and high energy density all-solid-state rechargeable (secondary) batteries [1-3, 6-9]. Furthermore, galvanic cells using solid electrolytes being employed for the determination of thermodynamic quantities such as Gibbs energies, entropies, enthalpies, and kinetic parameters such as chemical diffusion coefficients for fundamental scientific characterization / and understanding of materials [8-14].

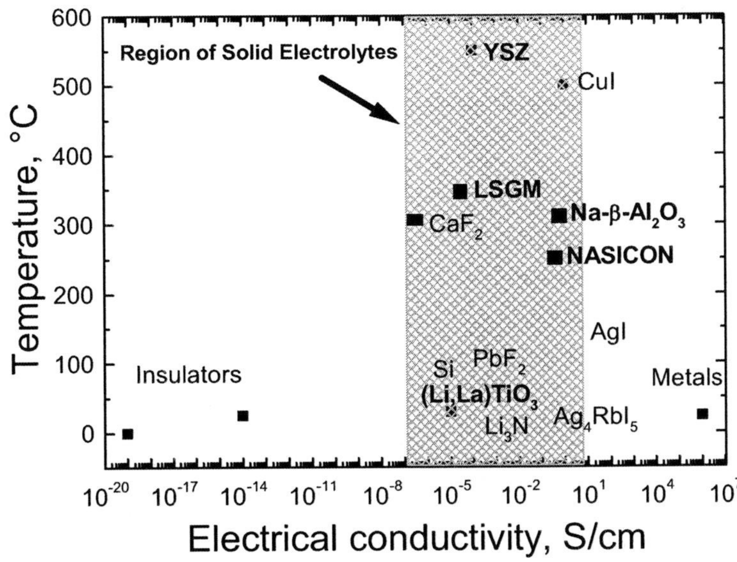

Figure 1. Electrical conductivities of typical solid electrolytes with metals, semiconductors and insulators and liquid electrolytes are compared at room temperature.

Usually, solid electrolytes are employed in practical applications in combination with predominantly electronically conducting electrodes. For example, Sr-doped $LaMnO_3$ and Ni,YSZ (Y_2O_3 stabilized ZrO_2) cermet should be chemically stable in contact with YSZ electrolyte during operation of SOFCs over a sufficiently long period of time (Fig. 2) [15]. This makes the development of ionic devices, especially those requiring large electrical currents, for example, electrical energy storage and conversion systems, to one of the most challenging but also attractive tasks for solid-state materials scientists.

2. Oxide Ion Conductors: General Materials Aspects

Ionic transport in fluorite-type structured zirconia (ZrO_2) was first reported by Nernst in 1897 [16]. However, it has taken nearly half a century to prove that not the small tetravalent Zr^{4+} ions (ionic radii: 0.79 Å) are conducting but the much larger divalent O^{2-} ions (ionic radii: 1.40 Å), which is apparently the result of weaker electrostatic interactions [17,18]. Subsequently, a large number of inorganic metal oxides possessing fluorite (AO_2), pyrochlore ($A_2B_2O_7$), perovskite (ABO_3), brownmillerite ($A_2B_2O_5$), scheelite (ABO_4), Aurivillius (A_2BO_6), layered perovskite-type (A_2BO_4) and apatite ($La_{10}(SiO_4)_6O_2$) structures were explored as oxide ion conductors (Fig. 3) [18-29]. Commonly, acceptor doping, i.e., by lower valent cations than that of A- or B-site cations is being employed to produce anion-deficiencies (oxygen stoichiometry less than that of the parent compound), which is the most essential requirement for oxide ion conduction. Table 1 lists typical examples of oxide ion electrolytes with each type of crystal structure [30-50].

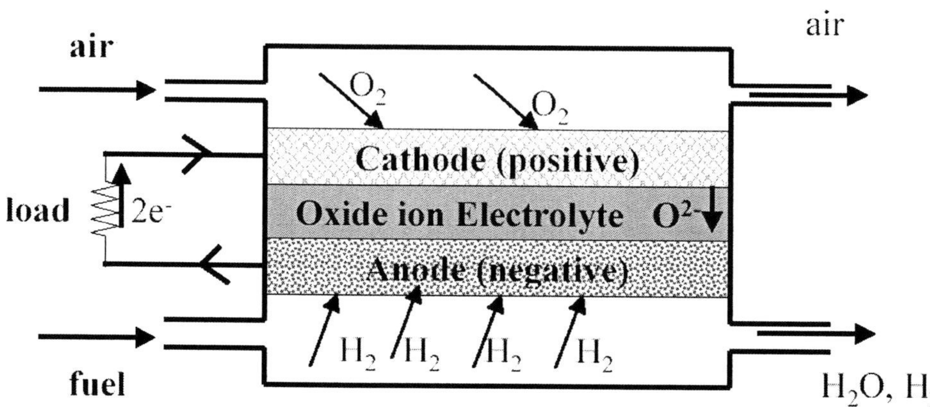

(a)

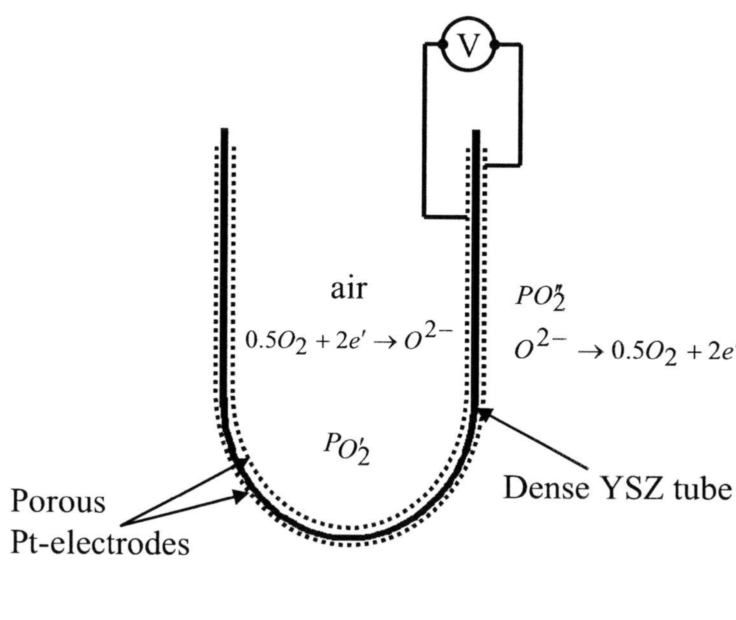

(b)

Figure 2. Schematic representations show the working principles of (a) solid oxide fuel cells and (b) oxygen sensors. In both cases, solid electrolyte is sand witched between predominate electronic conductors and expected to be stable against chemical reaction with electrodes and stable under the imposed oxygen potentials.

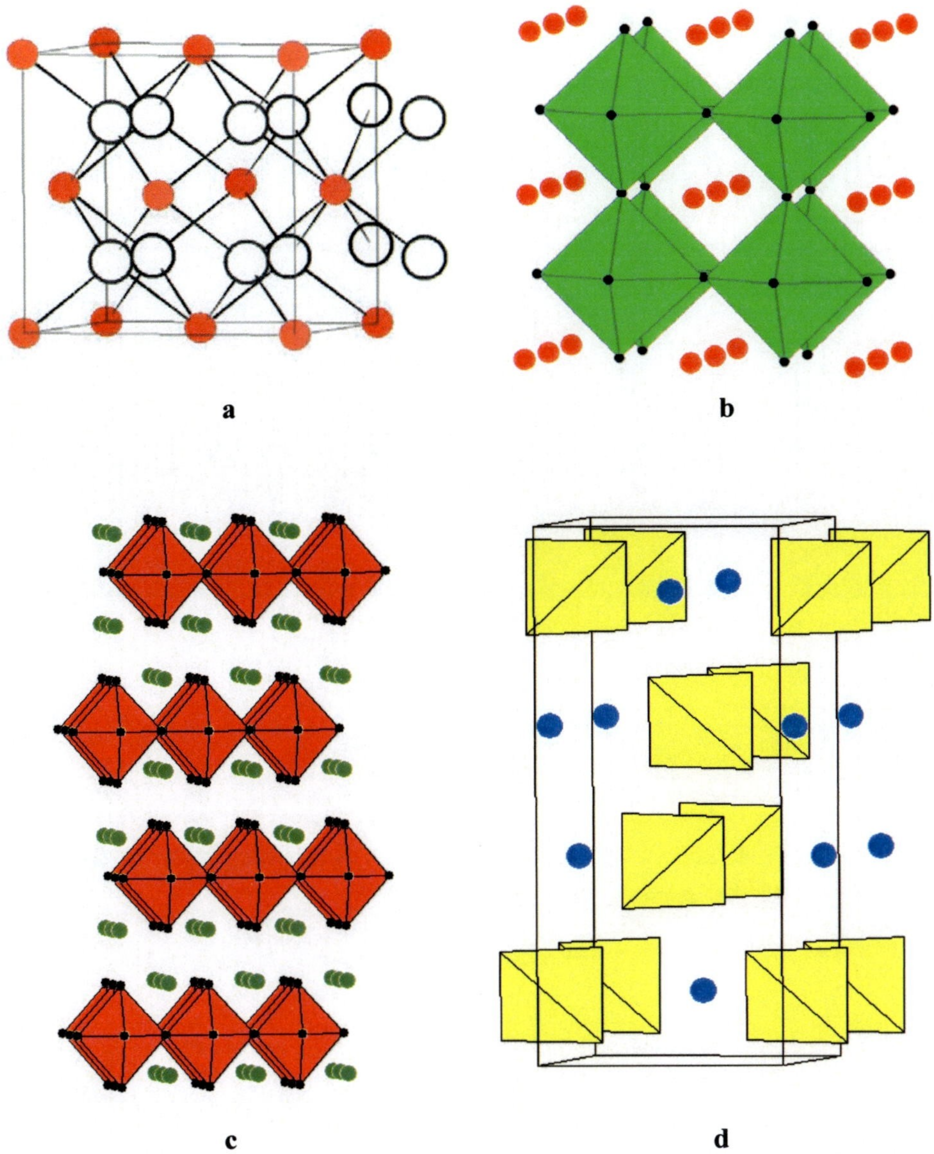

Figure 3. Crystal structures for fast oxide ion conduction: (a) fluorite (e.g., doped ZrO_2: Zr-closed spheres: O-open spheres), (b) perovskite (e.g., doped $SrTiO_3$: Sr-closed spheres; TiO_6: octahedra), (c) layered perovskite-type (e.g., doped Sr_2TiO_4: Sr-closed spheres; TiO_6: octahedral) (d) scheelite (e.g., $CaWO_4$: Ca-closed spheres; WO_4: tetrahedra), (e) brownmillerite (e.g., $Ba_2In_2O_5$: Ba-closed spheres; InO_6: octahedra, InO_4: tetrahedra), (f) Aurivillius (e.g., Bi_2WO_6: Bi_2O_2: tetrahedra; WO_6: octahedra), (g) pyrochlore (e.g., $Gd_2Ti_2O_7$: Gd-closed spheres; TiO_6: octahedra), and (h) apatite (e.g., $La_{10}(SiO_4)O_2$; O-large closed spheres; La-small closed spheres; SiO_4: tetrahedra) [2,6,8,11, 13,18-30].

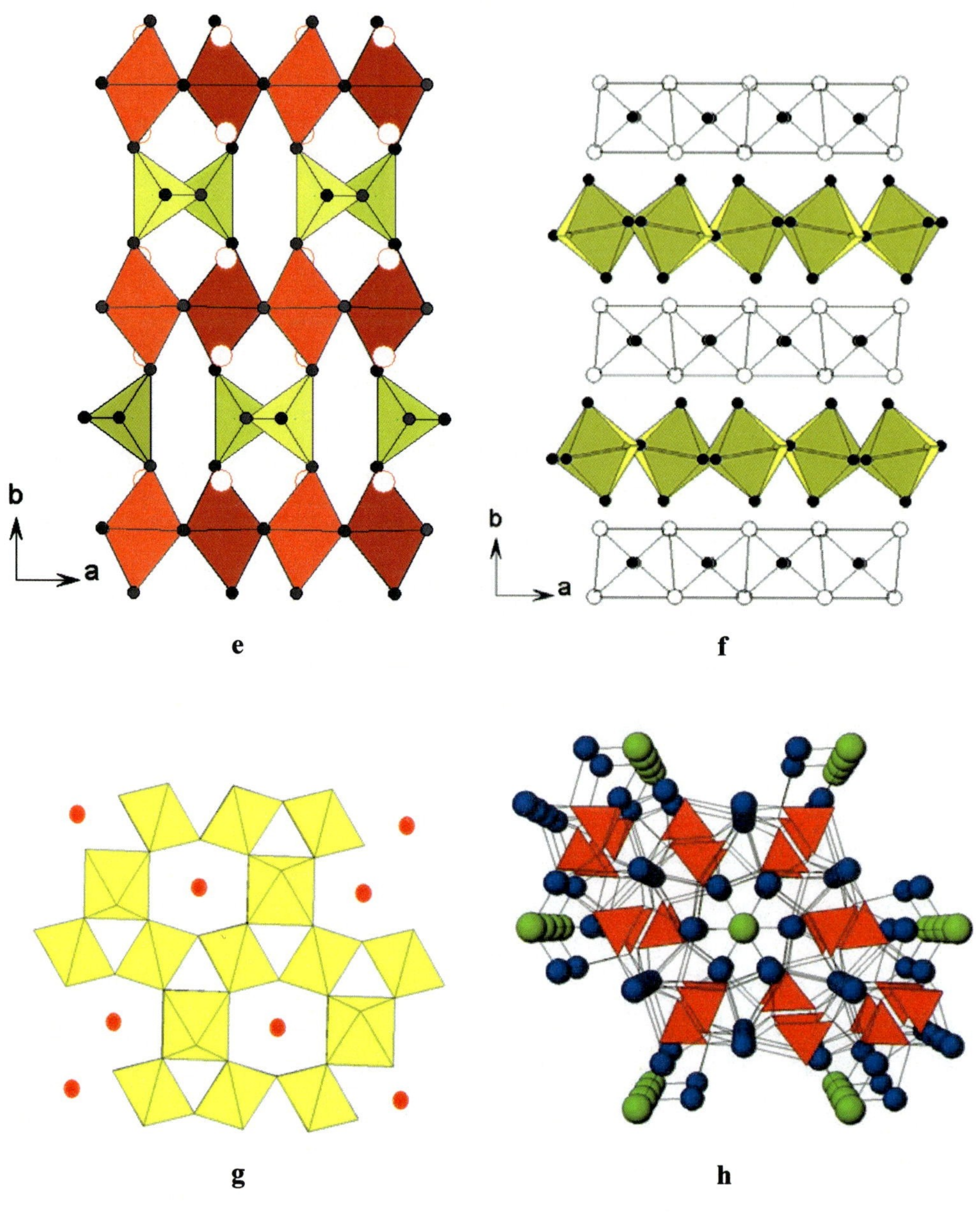

Figure 3. Crystal structures for fast oxide ion conduction (continued)

Figure 4 shows the Arrhenius plots for the oxide ion conductivity of the selected metal oxides. Among them, fluorite structured YSZ is extraordinarily chemically stable against chemical reaction with many electrode materials up to temperatures above 800 °C and electrochemically stable over a wide range of oxygen partial pressures ($1\text{-}10^{-40}$ atm) in the temperature range 700-1000 °C. Accordingly, YSZ is being employed as potential electrolyte, Ni-YSZ as negative electrode and Sr-doped $LaMnO_3$ as cathode for SOFCs, which operate efficiently at elevated temperatures [51-62].

Table 1. Structure-type, chemical composition and electrical conductivity data of selected oxide ion electrolytes

Structure-Type	Chemical Composition	$\sigma_{500\ °C}$ (S/cm)	$\sigma_{700\ °C}$ (S/cm)	E_a (eV)	Reference
Fluorite	Y_2O_3-ZrO_2 (YSZ)	2.5 x 10^{-4}	10^{-2}	0.98	[30]
	CaO-ZrO_2 (CSZ)	*< 10^{-5}*	1.0 x 10^{-3}	-	[30]
	$Ce_{0.9}Gd_{0.1}O_{2-\delta}$ (CGO)	9.5 x 10^{-3}	5.4 x 10^{-3}	0.64	[31]
	$Ce_{0.9}Sm_{0.1}O_{2-\delta}$ (CSO)	3.3 x 10^{-3}	2.0 x 10^{-3}	0.66	[31]
	$(Bi_2O_3)_{0.75}(Y_2O_3)_{0.25}$	8.0 x 10^{-3}	0.14	-	[20,21]
Pyrochlore	$Gd_2Ti_2O_7$	-	~10^{-3}	-	[32]
	$Gd_{1.9}Ca_{0.10}Ti_2O_{7-\delta}$	-	~10^{-2}	0.63	[32]
Perovskite	$CaTi_{0.95}Mg_{0.05}O_{3-\delta}$	*< 10^{-5}*	4.0 x 10^{-3}	-	[33]
	$La_{1-x}Sr_xGa_{1-y}Mg_yO_{3-\delta}$ (LSGM)	*< 10^{-3}*	1 x 10^{-1}	1.07	[34,35]
Brownmillerite	$Ba_2In_2O_5$	1.5 x 10^{-5}	5.0 x 10^{-3}	0.65	[36]
	$Ba_3In_2ZrO_8$	8.0 x 10^{-2}	-	0.60	[36]
	$Ba_2In_{1.5}Ga_{0.5}O_5$	5.0 x 10^{-4}	6.3 x 10^{-3}	1.04	[37]
Scheelite	$SrWO_4$	-	3.2 x 10^{-5}	0.40	[38]
	$Pb_{0.9}Sm_{0.1}WO_{4-\delta}$	3.2 x10^{-3}	2.1 x 10^{-2}	0.62	[39]
Aurivillius	$Bi_4V_2O_{11}$	1.0 x 10^{-2}	0.7	-	[40]
	$Bi_4V_{1.8}Cu_{0.2}O_{11-\delta}$	5.0 x 10^{-2}	0.1	-	[41,42]
	$Bi_4V_{1.8}Ti_{0.2}O_{11-\delta}$	5.0 x 10^{-2}	0.1	-	[41,42]
	$Bi_2Sr_2Nb_2AlO_{11.5}$	1.6 x 10^{-4}	8.0 x 10^{-3}	1.26	[43]
	$Bi_2Sr_2Nb_2GaO_{11.5}$	1.0 x 10^{-4}	3.2 x 10^{-3}	1.28	[43]
Apatite	$La_{10}Si_6O_{27}$	4.3 x 10^{-3}	1.1 x 10^{-2}	0.62	[44]
	$Pr_{10}Si_6O_{27}$	2.8 x 10^{-3}	6.0 x 10^{-3}	0.55	[44]
	$La_{10}(SiO_4)_6O_3$	1.8 x 10^{-4}	1.4 x 10^{-3}	0.69	[45]
	$Nd_{10}(SiO_4)_6O_3$	1.5 x 10^{-4}	8.4 x 10^{-4}	0.71	[45]
	$Sm_{10}(SiO_4)_6O_3$	1.4 x 10^{-5}	8.7 x 10^{-5}	0.79	[45]
	$La_{10}(SiO_4)_6O_3$	2.3 x 10^{-4}	1.0 x 10^{-3}	1.3	[46]
Cuspidine-Type	$La_4(Ga_2O_7)O_2$	*< 10^{-7}*	1.0 x 10^{-4}	-	[47]
Tysonite	LaO_xF_{3-2x} (x = 0.6)	3.2 x 10^{-2}	-	-	[48,49]
	$La_{0.9}Sr_{0.1}O_{0.5}F_{1.9}$	3.2 x 10^{-4}	1.0 x 10^{-2}	-	[49]
Miscellaneous	$La_{1.5}GeO_{5-\delta}$	*3.0 x 10^{-4}*	1.0 x 10^{-1}	0.86	[50a]
	$La_{1.5}Sr_{0.17}GeO_5$-δ	*1.0 x 10^{-3}*	3.2 x 10^{-2}	-	[50a]
	$La_2Mo_2O_9$	5.0 x 10^{-5}	3.0 x 10^{-2}	1.2	[50b]

The values in italics are extrapolated from the Arrhenius plots.

The high operation temperature of current SOFCs results in several materials problems, which may prevent commercialization. They are (i) interfacial diffusion of Sr and La between the electrode and electrolyte that leads to the formation of poor conducting reaction products, e.g., $SrZrO_3$ and $La_2Zr_2O_7$; (ii) mechanical stress due to different thermal expansion coefficients (TEC), which destroys good electrolyte and electrode contacts; (iii) high cost of the bi-polar separators which are required at high temperatures for long-term stability and

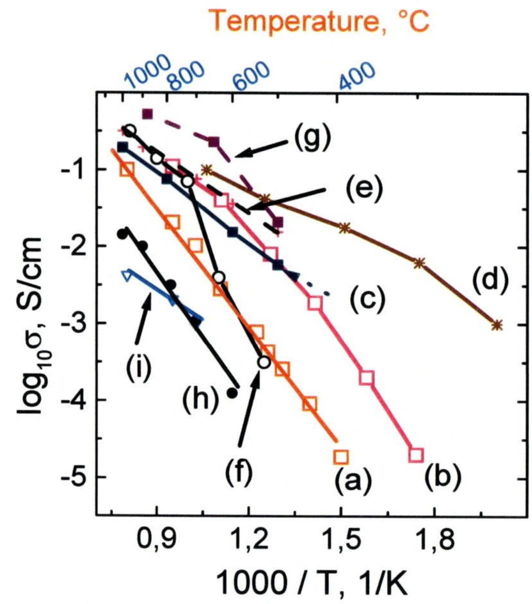

Figure 4. Arrhenius plots for oxide ion conduction of selected metal oxides (a) Y_2O_3-stabilized ZrO_2 (YSZ), (b) $La_{0.9}Sr_{0.1}Ga_{0.8}Mg_{0.2}O_{3-\delta}$ (LSGM), (c) Sc_2O_3-stabilized ZrO_2, (d) $Bi_4V_{1.8}Cu_{0.2}O_{11-\delta}$, (e) Gd-doped CeO_2, (f) La_2GeO_5, (g) Eu-doped Bi_2O_3, (h) Mg-doped $CaTiO_3$ and (i) $Gd_2Ti_2O_7$.

stability in oxidizing and reducing conditions and (iv) lack of appropriate sealing materials which are chemically inert against reaction with all components of fuel cells [63-69]. Accordingly, the current research work is directed mainly to find new oxide ion electrolytes for low-temperature application (< 600 °C), to decrease the operating temperature and achieve long-term stable SOFCs. Furthermore, low operation temperature has great advantages over the presently mostly employed high temperature (> 900 °C) such as flexibility in the choice of the electrodes, bi-polar materials, glass ceramic seals as well as long-life time of the devices.

Recent discoveries of fast oxide ion conductors by Ishihara et. al [34] and Goodenough et. al [35] were based on $LaGaO_3$ ($\cong ABO_3$) perovskites by substitutions of La by Sr and Ga by Mg which turned out to be an alternative to the YSZ electrolyte since a member of the series, $La_{1-x}Sr_xGa_{1-y}Mg_yO_{3-\delta}$ (LSGM); x = 0.1, y = 0.2, $La_{0.9}Sr_{0.1}Ga_{0.8}Mg_{0.2}O_{2.85}$, exhibits oxide ion conductivity of 0.1 S/cm at 800 °C which is about three times higher than that of YSZ at 800 °C. Rare earth (RE)-doped CeO_2 exhibits predominantly ionic conductivity in the temperature range 400-600 °C under oxygen partial pressures of 1 - 10^{-19} atm, which is sufficient to employ this compound as electrolyte in fuel cells with an appropriate choice of the fuel gases. Accordingly, RE-doped CeO_2 is presently accepted as solid electrolyte for IT-SOFCs. Employing $Ce_{0.8}Sm_{0.2}O_{1.9}$ (CSO) and $Ce_{0.8}Gd_{0.2}O_{1.9}$ (CGO) as electrolyte and using hydrocarbon as fuel and air as oxidant gas show a better performance than YSZ and LSGM at 550 °C [70-72]. Our research work is mainly focused on perovskite-like, fluorite-type and scheelite-related structures to understand the chemical composition- structure-electrical properties relationships, which will be reviewed.

3. Investigations on ABO_3 Perovskite-Like Structure Metal Oxides

The existence of oxide ion conduction in perovskite structured materials was first reported for acceptor doped $CaTiO_3$ and $LaAlO_3$ by Takahashi and Iwahara in 1971 [33]. However, research and development of oxide ion electrolytes with perovskite and its related structure for the possible application in SOFCs and oxygen sensors has only been performed in recent years. For example, partial substitution of Nd by Ca and Al by Ga in $NdAlO_3$ yields an anion deficient perovskite, $Nd_{0.9}Ca_{0.1}Al_{0.5}Ga_{0.5}O_{3-\delta}$, which exhibits oxide ion conductivity of 10^{-2} S/cm at 700 °C [73]. Solid solution $CaTi_{1-x}M_xO_{3-\delta}$ (M = Ga, Sc) in the x range of 0.1-0.25 show oxide ion conductivity comparable to that of YSZ in the temperature regime 800-1000 °C [74].

Due to high ionic conductivity in LSGM perovskites, a large number of studies have been performed to understand the composition-structure-electrical properties [75-82]. Replacement of trivalent Ga^{3+} by other trivalent ions such as Al^{3+}, Sc^{3+} and In^{3+} decreases the electrical conductivity [77]. Among the alkaline earth ions occupying the A-site, Sr provides the highest conductivity compared to Ca and Ba-substituted compounds. Furthermore, the divalent Mg in LSGM has been replaced as well completely as partially by other divalent and trivalent transition metal ions that include all first row transition metals. The complete replacement of Mg by V, Cr, Mn, Fe, Co, and Ni in LSGM yields novel mixed oxide ion and electronic conductors [78], which may find application in SOFCs as cathode materials. Partial replacement of Co (0.05 mol.-%) for Mg in LSGM, forming the composition $La_{0.9}Sr_{0.1}Ga_{0.8}Mg_{0.15}Co_{0.05}O_{2.85}$, increases slightly the ionic conductivity of LSGM [79].

It has been observed that LSGM suffers from exposure to CO and CO_2 atmospheres due to the formation of alkaline carbonates and high Ga volatilization in reducing atmospheres at elevated temperatures [80-82] and form Pt-Ga alloys when Pt electrodes are being employed, similar to the formation of Pt-Zr alloys in the case of zirconia electrolytes in contact with Pt electrodes under reducing conditions [83]. Accordingly, further understanding and improving the LSGM perovskites with regard to chemical stability, ionic conductivity by appropriate substitutions of A-(La,Sr) and B-(Ga,Mg) site cations are important objectives among the materials scientists. Replacement of the costly and less stable Ga by other suitable metal ions without affecting the electrical properties in LSGM and identification of Ga free perovskite-type and perovskite-related structured materials remains challenging.

3.1. Doped LSGM and LSIM Perovskites

Perovskite-type oxides of the nominal chemical compositions $La_{0.9}K_{0.1}Ga_{0.9}Mg_{0.1}O_{2.85}$, $La_{0.9}Sr_{0.1}Ga_{0.7}Al_{0.1}Mg_{0.2}O_{2.85}$, $La_{0.9}Sr_{0.1}Ga_{0.7}In_{0.1}Mg_{0.2}O_{2.85}$, $La_{0.45}Pr_{0.45}Sr_{0.1}Ga_{0.8}Mg_{0.2}O_{2.85}$, $La_{0.4}Pr_{0.4}Sr_{0.2}Ga_{0.8}Mg_{0.2}O_{2.8}$, $La_{0.9}Sr_{0.1}In_{0.8}Mg_{0.2}O_{2.85}$, $La_{0.45}Pr_{0.45}Sr_{0.1}In_{0.8}Mg_{0.2}O_{2.85}$ and $La_{0.4}Pr_{0.4}Sr_{0.2}In_{0.8}Mg_{0.2}O_{2.8}$ have been prepared by solid-state reactions at elevated temperatures (1000-1400 °C) using appropriate mixtures of K_2CO_3, $SrCO_3$, La_2O_3 (dried at 900°C over night), Pr_6O_{11}, Al_2O_3, Ga_2O_3 and In_2O_3 [84]. The powder XRD data (Siemens D-5000 or Seifert 3000, CuK_α) (Figures 5 and 6) show the formation of single-phase perovskite-type materials. Both K- and Pr- substitutions for La/Sr-site, and In-substitution for Ga-site in

LSGM result in simple cubic perovskite structure, while LSIM and doped LSIM crystallize in an orthorhombic $GdFeO_3$ type structure. Table 2 lists the lattice parameters of synthesized perovskite-like compounds. The formation of ordered and distorted $GdFeO_3$ perovskites of In-compounds is due to the smaller tolerance factor (t) value as compared to that of LSGM. The tolerance factor can be calculated using the formula:

$$t = \frac{r_A + r_O}{\sqrt{2}(r_B + r_O)} \tag{1}$$

where r_A, r_B and r_O represent radii of A-site cations, B-site cations and oxygen ions, respectively. The tolerance factor for $La_{0.9}Sr_{0.1}Ga_{0.8}Mg_{0.2}O_{2.85}$ (LSGM) and $La_{0.9}Sr_{0.1}In_{0.8}Mg_{0.2}O_{2.85}$ (LSIM) is 0.90 and 0.84, respectively.

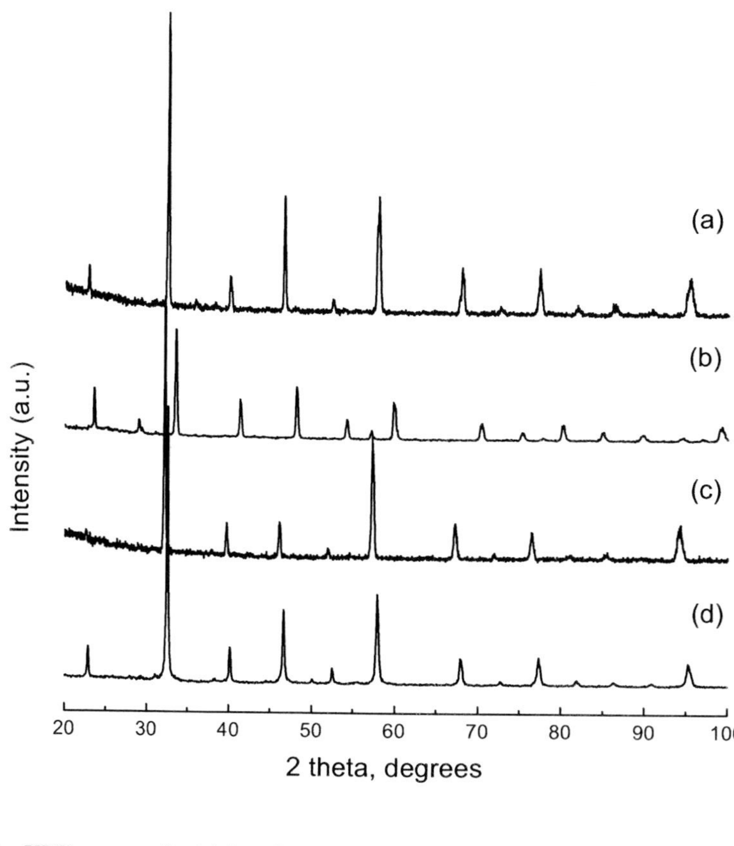

Figure 5. Powder XRD patterns for (a) $La_{0.9}Sr_{0.1}Ga_{0.8}Mg_{0.2}O_{3-\delta}$ (LSGM), (b) $La_{0.9}Sr_{0.1}Ga_{0.7}Al_{0.1}Mg_{0.2}O_{3-\delta}$ (Al-LSGM), (c) $La_{0.9}Sr_{0.1}Ga_{0.7}In_{0.1}Mg_{0.2}O_{3-\delta}$ (In-LSGM) and (d) $La_{0.9}K_{0.1}Ga_{0.8}Mg_{0.2}O_{3-\delta}$ (LKGM) [84].

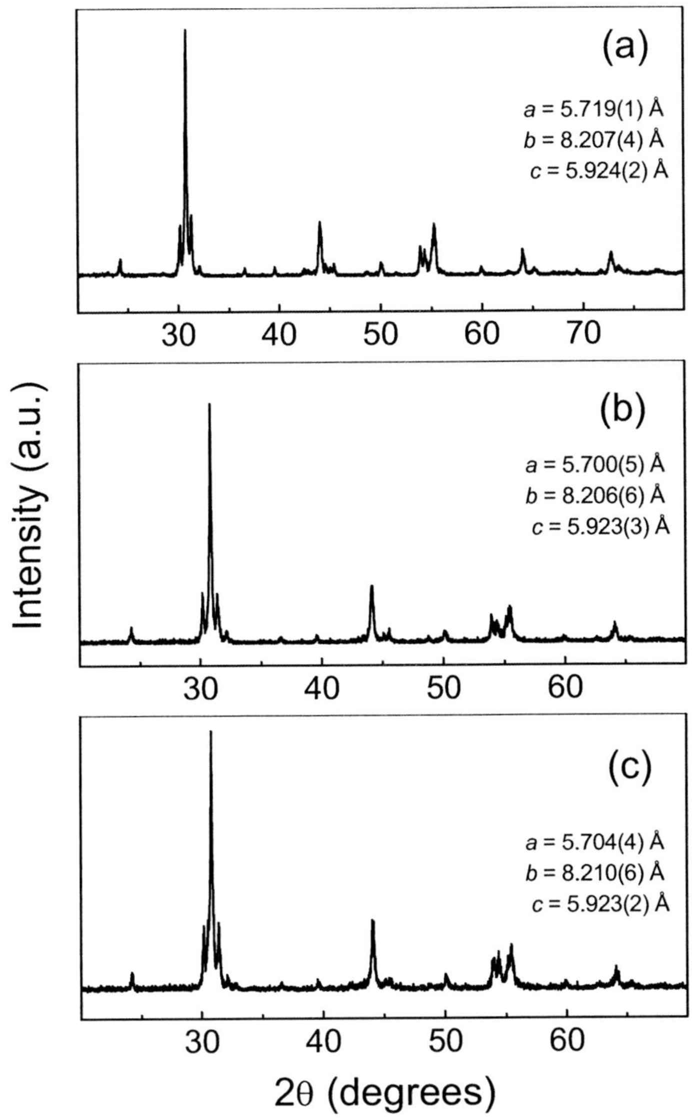

Figure 6. Powder XRD patterns of (a) $La_{0.9}Sr_{0.1}In_{0.8}Mg_{0.2}O_{3-\delta}$ (LSIM) (b) $Pr_{0.45}La_{0.45}Sr_{0.1}In_{0.8}Mg_{0.2}O_{3-\delta}$ (Pr-LSIM) and (c) $Pr_{0.40}La_{0.40}Sr_{0.2}In_{0.8}Mg_{0.2}O_{3-\delta}$ ($Pr_{0.40}$-LSIM) [84].

Figure 7 shows a typical impedance plot (5Hz-13 MHz; HP 4192 A impedance and gain phase analyzer) obtained in air for LSGM at 254 °C. We see the bulk contribution at high frequency and grain-boundary resistance at the low frequency side. The magnitude of grain boundary conductivity is slightly higher than that of the bulk conductivity. The electrical conductivity of LSGM samples with three different cell-constants exhibits similar bulk and grain-boundary conductivities (Fig. 8) [84b] and is comparable to that of the literature data [75]. Replacement of Sr by K in LSGM shows lower electrical conductivity (8.56×10^{-3} S/cm at 700 °C) with high activation energy of 1.42 eV compared to that of LSGM (1.07 eV) [34].

Table 2. Chemical composition, electrical conductivity and lattice parameters of investigated oxide ion conducting perovskites

Compounds	$\sigma_{(700\,°C)}$ (S/cm)	$\sigma_{(200\,°C)}$ (S/cm)	E_a (eV)	Lattice parameters (Å)		
				a	b	c
$La_{0.9}K_{0.1}Ga_{0.9}Mg_{0.1}O_{2.85}$	8.56×10^{-3}	1.49×10^{-10}	1.42	3.896(3)	-	-
$La_{0.9}Sr_{0.1}Ga_{0.7}Al_{0.1}Mg_{0.2}O_{2.85}$	2.81×10^{-4}	5.19×10^{-11}	1.23	5.347(5)	7.559(5)	5.354(8)
$La_{0.9}Sr_{0.1}Ga_{0.7}In_{0.1}Mg_{0.2}O_{2.85}$	2.36×10^{-2}	1.25×10^{-8}	1.09	3.933(3)	-	-
$La_{0.45}Pr_{0.45}Sr_{0.1}Ga_{0.8}Mg_{0.2}O_{2.85}$	2.96×10^{-2}	2.89×10^{-7}	0.91	3.888(2)	-	-
$La_{0.4}Pr_{0.4}Sr_{0.2}Ga_{0.8}Mg_{0.2}O_{2.8}$	3.00×10^{-2}	2.89×10^{-7}	0.91	3.898(1)	8.198	11.796
$La_{0.9}Sr_{0.1}In_{0.8}Mg_{0.2}O_{2.85}$	2.48×10^{-3}	6.67×10^{-8}	0.83	5.719(1)	8.206(6)	5.923(3)
$La_{0.45}Pr_{0.45}Sr_{0.1}In_{0.8}Mg_{0.2}O_{2.85}$	2.48×10^{-3}	9.70×10^{-8}	0.84	5.700(5)	8.210(4)	5.923(2)
$La_{0.4}Pr_{0.4}Sr_{0.2}In_{0.8}Mg_{0.2}O_{2.8}$	9.48×10^{-3}	3.17×10^{-5}	0.44	5.704(4)	7.559(5)	5.354(8)

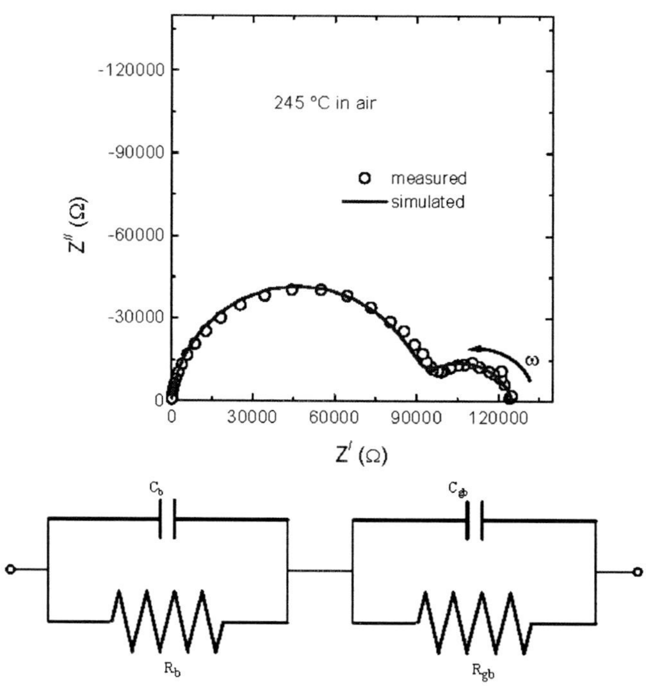

Figure 7. (a) Typical impedance plot obtained in air for $La_{0.9}Sr_{0.1}Ga_{0.8}Mg_{0.2}O_{3-\delta}$ (LSGM). The bulk resistance at high frequency and grain-boundary contribution at low frequency side were observed. The line passing though the data points represents simulated data using the EQUIVALENT circuit consisting of (b) two parallel resistances-capacitances (RQ)(RQ) [84].

Partial substitution of Ga by Al or In in LSGM decreases the electrical conductivity. The decrease in electrical conductivity may be attributed to the change in the crystal structure and lattice parameters.

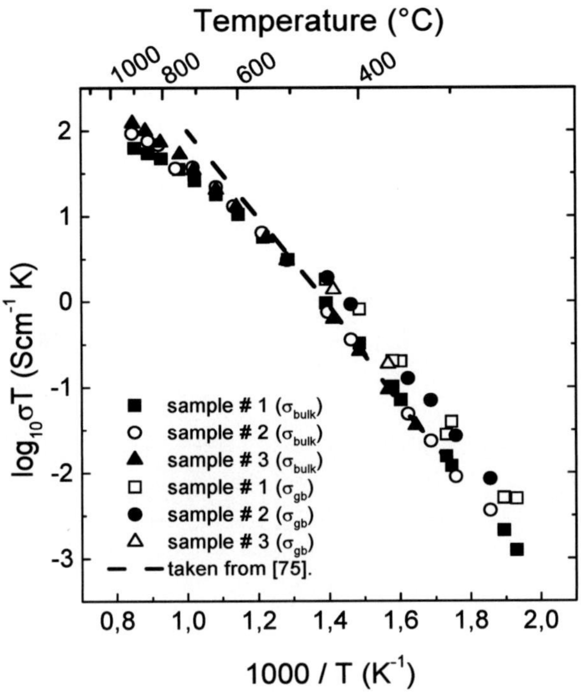

Figure 8. The Arrhenius plots for the bulk (σ_{bulk}), grain-boundary (σ_{gb}) conductivity of $La_{0.8}Sr_{0.2}Ga_{0.8}Mg_{0.2}O_{2.80}$ (LSGM) samples with three different cell constants (sample # 1: 1.28 mm thickness; 8.23 mm diameter; sample # 2: 2.90 mm thickness; 8.84 mm diameter; sample # 3: 5.64 mm thickness; 9.04 mm diameter) [84b]. The conductivity data of LSGM from literature is shown for comparison [75]. The data obtained during the heating and cooling cycle follow the same line, indicating the equilibrium conductivity values.

Unlike other rare earth substitutions of LSGM, substitution of La by Pr in LSGM does not affect the electrical conductivity, except for a small decrease at high temperature, while at low temperatures the conductivity corresponds to that of LSGM. The activation energy for the electrical conductivity (0.91 eV) is found to be comparable to those of the literature values. Both $La_{0.45}Pr_{0.45}Sr_{0.1}Ga_{0.8}Mg_{0.2}O_{2.85}$ and $La_{0.4}Pr_{0.4}Sr_{0.2}Ga_{0.8}Mg_{0.2}O_{2.8}$ exhibit similar conductivity values [84]. Among the investigated oxides in this study, $La_{0.4}Pr_{0.4}Sr_{0.2}In_{0.8}Mg_{0.2}O_{2.8}$, exhibits the highest electrical conductivity of 3.2×10^{-5} S/cm at 200 °C with the lowest activation energy of 0.44 eV [84]. The increase in the electrical conductivity may be due to the higher number of oxygen vacancies or the increase in the electronic conductivity due to mixed valent Pr doping (Fig. 9).

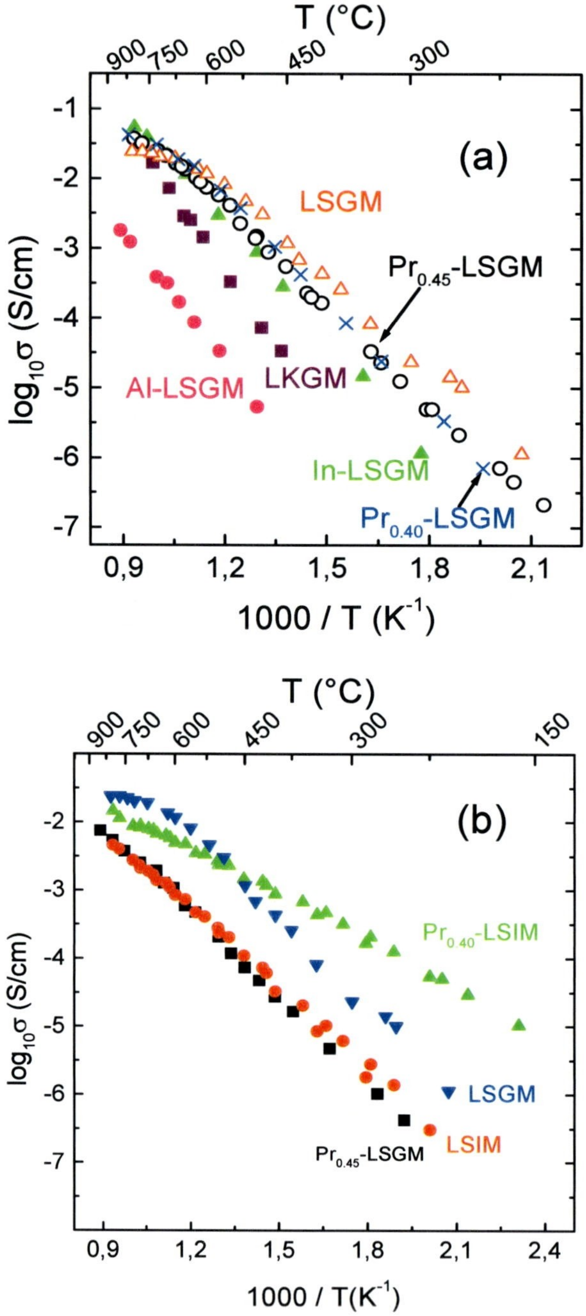

Figure 9. The Arrhenius plots for total (bulk (σ_{bulk}) + grain-boundary (σ_{gb})) conductivity of (a) LSGM [$La_{0.9}Sr_{0.1}Ga_{0.8}Mg_{0.2}O_{2.85}$ (LSGM), $La_{0.45}Pr_{0.45}Sr_{0.1}Ga_{0.8}Mg_{0.2}O_{2.85}$ ($Pr_{0.45}$-LSGM), $La_{0.40}Pr_{0.40}Sr_{0.2}Ga_{0.8}Mg_{0.2}O_{2.80}$ ($Pr_{0.40}$-LSGM), $La_{0.9}Sr_{0.1}Ga_{0.7}Al_{0.1}Mg_{0.2}O_{2.85}$ (Al-LSGM), $La_{0.9}Sr_{0.1}Ga_{0.7}In_{0.1}Mg_{0.2}O_{2.85}$ (In-LSGM), $La_{0.9}K_{0.1}Ga_{0.8}Mg_{0.2}O_{2.85}$ (K-LSGM)] and (b) LSIM [$La_{0.9}Sr_{0.1}In_{0.8}Mg_{0.2}O_{2.85}$ (LSIM), $La_{0.45}Pr_{0.45}Sr_{0.1}In_{0.8}Mg_{0.2}O_{2.85}$ ($Pr_{0.45}$-LSIM), $La_{0.40}Pr_{0.40}Sr_{0.2}In_{0.8}Mg_{0.2}O_{2.80}$ ($Pr_{0.40}$-LSIM)] based perovskites [84]. The data obtained during the heating and cooling cycle follow the same line, indicating the equilibrium conductivity values.

Oxygen partial pressure dependent electrical measurements of $La_{0.9}Sr_{0.1}In_{0.8}Mg_{0.2}O_{2.85}$ (LSIM) show p-type electronic conduction at high oxygen partial pressure in the range 10^{-6}-0.21 atm at 750 °C with a slope of 1/6 (Fig. 10). This behaviour is assumed to be due to the following defect reaction [85]:

$$2V_O^{\bullet\bullet} + O_2 \rightarrow 2O_O + 4h^{\bullet} \tag{2}$$

It is interesting to note that the electrical conductivity of LSIM does not vary significantly on the oxygen partial pressure regime 10^{-5}- 10^{-22} atm, indicating that it may be an ionic conductor. A similar p-type electronic conduction has been observed for other $La_{0.45}Pr_{0.45}Sr_{0.1}In_{0.8}Mg_{0.2}O_{2.85}$, $La_{0.40}Pr_{0.40}Sr_{0.2}In_{0.8}Mg_{0.2}O_{2.85}$, and $La_{0.45}Pr_{0.45}Sr_{0.1}Ga_{0.8}Mg_{0.2}O_{2.85}$ materials. The average transference number ($\bar{t}_{O^{2-}}$) for ionic conduction of $La_{0.45}Pr_{0.45}Sr_{0.1}Ga_{0.8}Mg_{0.2}O_{2.85}$ can be obtained using the equation [86]:

$$\bar{t}_{O^{2-}} = \frac{EMF_{measured}}{EMF_{calculated}} \tag{3}$$

(where $E_{measured}$ and $E_{calculated}$ represent measured and thermodynamic EMF, respectively) for the galvanic cell (I) employing air and hydrogen with $La_{0.45}Pr_{0.45}Sr_{0.1}Ga_{0.8}Mg_{0.2}O_{2.85}$ as an electrolyte:

(-) H_2, Pt || $La_{0.45}Pr_{0.45}Sr_{0.1}Ga_{0.8}Mg_{0.2}O_{2.85}$ || Pt, air (+) (I)

The transference number for oxide ion conduction is found to be in the range 0.78- 0.90 at 450-650 °C (Fig. 11). Accordingly, we believe that substitution of Pr for La in LSGM results in mixed ionic-electronic conduction. Powder XRD data reveal that Pr-doped LSGM was found to be structurally stable at low oxygen partial pressure of 10^{-22} atm (obtained by mixing CO_2 and H_2 gas) at 750 °C for 48 h, but, the corresponding In samples are not stable. We see several new peaks corresponding to La_2SrO_x, $La_2Sr_2O_5$ and La_4SrO_7 impurity phases in the LSIM based perovskites. The above results indicate that LSGM based perovskites are more stable than that of the In-analogue.

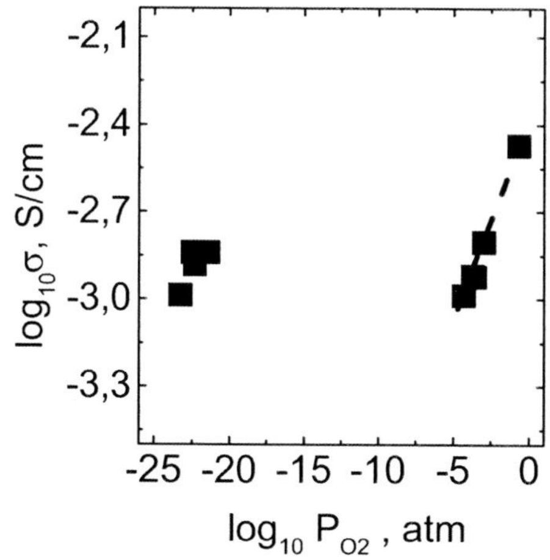

Figure 10. Electrical conductivity as a function of oxygen partial pressure at 750 °C of $La_{0.9}Sr_{0.1}In_{0.8}Mg_{0.2}O_{2.85}$ (LSIM). The broken line passing through the data points is guide to the eye.

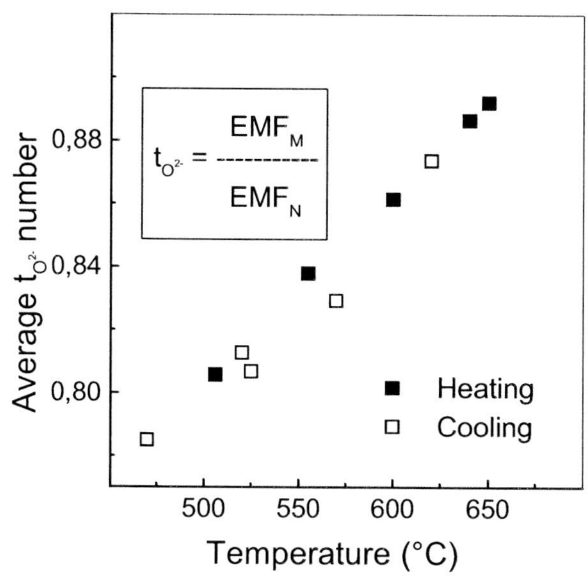

Figure 11. Average oxygen transference number for $La_{0.45}Pr_{0.45}Sr_{0.1}Ga_{0.8}Mg_{0.2}O_{2.85}$ in the temperature range 450-650 °C [84].

3.2. Perovskite-Type Solid Solution $SrSn_{1-x}Fe_xO_3$

Oxide ion conductors are generally approached on the basis of electronic insulators with large numbers of intrinsic or extrinsic ionic defects for fast ion transport [1,2,4,10,11,13]. Starting with a predominantly semiconducting material, we have modified the composition in such a way that the material becomes predominantly ionically conducting at intermediate activities of the mobile component and remains electronically conducting near the interfaces with the electrodes. We show that the nature of the conductivity of $SrSn_{1-x}Fe_xO_{3-\delta}$ changes with the oxygen activity of the material. $SrSn_{1-x}Fe_xO_{3-\delta}$ perovskites with low Fe-content exhibit both ionic and electronic conduction, while a higher Fe content (≥ 0.5) results in predominantly electronic conductivity over a wide range of oxygen partial pressures. The described material is likely only the first example of a large class of materials, which commonly appear to be semiconductors but are ionic conductors in a certain activity regime. This property has often not been recognized so far because activity dependant conductivity measurements have not been carried out. These activity dependent mixed conducting materials will have large practical advantages, especially for fast electrode kinetics since fast equilibration and high Wagner factors depend strongly on the presence of mobile electronic charge carriers [87].

$SrSn_{1-x}Fe_xO_{3-\delta}$ ($0 < x < 1$) perovskite oxides have been prepared by solid-state reactions of stoichiometric mixtures of $SrCO_3$, SnO_2 and Fe $(CO_2)_2.2H_2O$ at 1000 °C for 24 h and 1300 °C for 24 h with intermittent grinding and pelletizing using isostatic pressures [88-91]. Power XRD data reveal the formation of the perovskite-like $GdFeO_3$-type structure for x = 0 to 0.3, a simple cubic structure for x = 0.4 to 0.9 and orthorhombic structure for x = 1 members of $SrSn_{1-x}Fe_xO_{3-\delta}$ (Table 3) [91]. The lattice parameters of Fe-substituted $SrSnO_3$ decreases with increasing Fe content, suggesting that larger Sn^{4+} ions (0.69 Å) are replaced by smaller Fe^{3+} / Fe^{4+} ions (0.65 Å) (Fig. 12) [92]. The formation of $GdFeO_3$-type perovskites up to x = 0.3 indicates that both Sn and Fe are distributed over the octahedral sites of the distorted perovskite structure. We see that the x = 0.4-0.9 members crystallize in simple cubic structure suggesting that both Sn and Fe are disordered at perovskite B-site. The x = 1 member of the series, $SrFeO_{3-\delta}$, crystallizes in orthorhombic structure with a = 10.933(6), b = 7.7121(5) and c = 5.4733(5) Å [91].

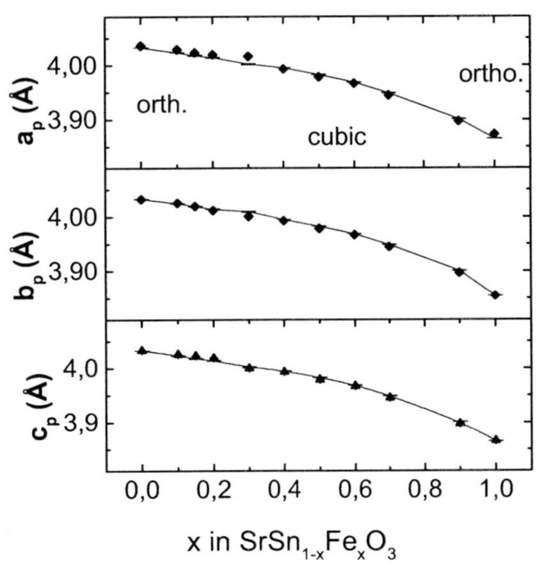

Figure 12. Lattice parameters as a function of the Fe content in $SrSn_{1-x}Fe_xO_{3-\delta}$ ($0 < x < 1$) [91].

Table 3. Electrical conductivity data of $SrSn_{1-x}Fe_xO_3$ perovskites

Perovskite	Lattice parameters (Å)			$\sigma_{25\,°C}$ (S/cm)	$\sigma_{700\,°C}$ (S/cm)	E_a (eV)
	a	b	c			
$SrSnO_3$	5.7082(3)	5.7035(3)	8.0659(6)	1.0×10^{-9}	1.0×10^{-9}	0.55
$SrSn_{0.9}Fe_{0.1}O_{3-\delta}$	5.6987(4)	5.6937(8)	8.0509(8)	1.0×10^{-9}	1.0×10^{-9}	0.55
$SrSn_{0.85}Fe_{0.15}O_{3-\delta}$	5.690(1)	5.685(1)	8.045(1)	8.4×10^{-9}	8.4×10^{-9}	0.56
$SrSn_{0.8}Fe_{0.2}O_{3-\delta}$	5.685(1)	5.674(2)	8.038(2)	1.8×10^{-7}	1.8×10^{-7}	0.53
$SrSn_{0.7}Fe_{0.3}O_{3-\delta}$	5.680(1)	5.659(1)	8.000(2)	2.9×10^{-6}	2.9×10^{-6}	0.48
$SrSn_{0.6}Fe_{0.4}O_{3-\delta}$	3.9934(1)	-	-	2.4×10^{-5}	2.4×10^{-5}	0.41
$SrSn_{0.5}Fe_{0.5}O_{3-\delta}$	3.9788(2)	-	-	3.6×10^{-4}	3.6×10^{-4}	0.40
$SrSn_{0.4}Fe_{0.6}O_{3-\delta}$	3.967(1)	-	-	3.3×10^{-3}	3.3×10^{-3}	0.19
$SrSn_{0.3}Fe_{0.7}O_{3-\delta}$	3.9450(2)	-	-	3.3×10^{-2}	3.3×10^{-2}	0.18
$SrSn_{0.1}Fe_{0.9}O_{3-\delta}$	3.8973(2)	-	-	1.1×10^{-1}	1.1×10^{-1}	0.12
$SrFeO_{3-\delta}$	10.9715(8)	7.7121(5)	5.4733(5)	8.6×10^{-1}	8.6×10^{-1}	-

AC impedance (5 Hz-13MHz) studies show that $SrFe_{0.1}Sn_{0.9}O_{3-\delta}$ exhibits two semicircles (which may be attributed to bulk and grain boundary contributions), while samples with higher Fe-content show only single semicircles or a part of the semicircles contribution [89]. There were no low-frequency tails indicating that the electrolyte-electrode interfaces were not blocking [93]. Accordingly, the low-frequency intercept corresponds to the electronic conductivity. Interestingly, the AC conductivity values [89] obtained in the present work for x up to 0.7 are in agreement with the DC values reported by Roh *et al* [88]. For example, the x = 0.5 member exhibits a conductivity of 3.65×10^{-4} S/cm at 25 °C, which is comparable to the DC value (~ 1.0×10^{-5} S/cm at 25 °C) [88]. However, there is a difference between the present value and reported value for compounds with high Fe content. The conductivity increases with increasing Fe content and is comparable to that of electronic conductors based on the perovskite structure (Figs. 13 and 14).

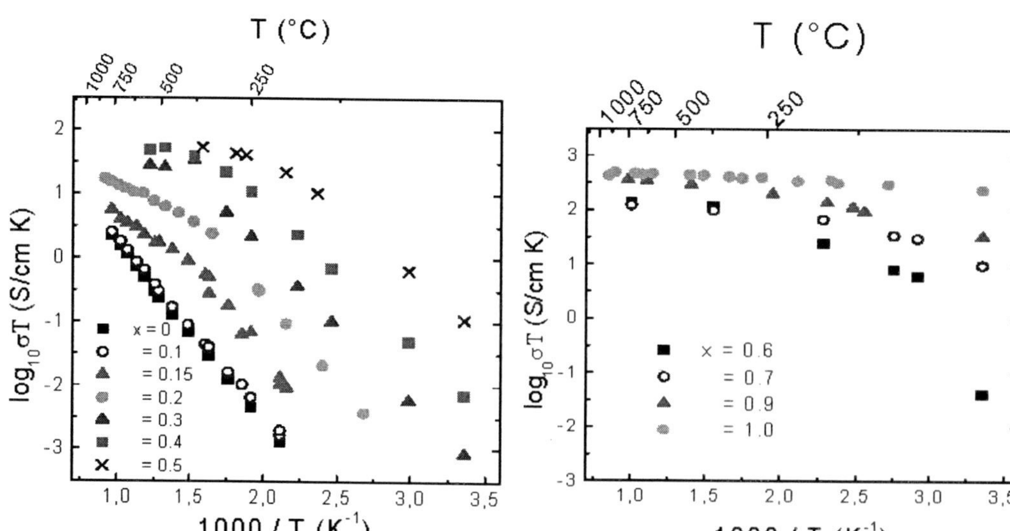

Figure 13. The Arrhenius plots for total electrical conductivity of $SrSn_{1-x}Fe_xO_{3-\delta}$ (0 < x < 1) in air [90].

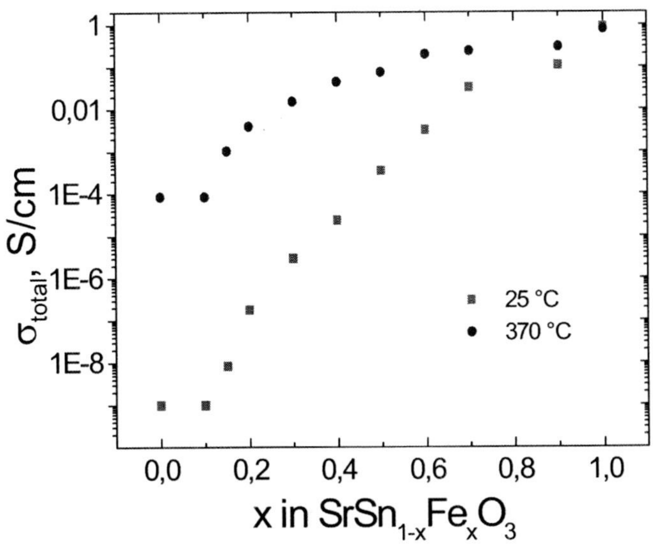

Figure 14. Variation of the total electrical conductivities as a function of the Fe content in $SrSn_{1-x}Fe_xO_{3-\delta}$ ($0 < x < 1$) at 25 and 370 °C in air [90].

Solid solution $SrSn_{1-x}Fe_xO_3$ containing low Fe content (< 0.3) exhibits regimes of predominant ionic and electronic conductivity (Fig. 15). At high and low oxygen partial pressures, for example, $SrSn_{0.9}Fe_{0.1}O_3$ exhibits p-and n-type electronic conductivity with a slope of ± 1/4 respectively. This behaviour is assumed to be due to the following defect reactions [85]:

$$V_O^{\bullet\bullet} + 0.5O_2 \leftrightarrow O_O^x + 2h^{\bullet} \quad (4)$$

$$O_O^x \leftrightarrow 0.5O_2 + V_O^{\bullet\bullet} + 2e' \quad (5)$$

at high and low oxygen partial pressures, respectively, assuming constant oxygen vacancy concentration. $O_O^x, V_O^{\bullet\bullet}, h^{\bullet}$ and e' represent regular lattice oxide ions, oxide ion vacancies, holes and electrons in the crystal, respectively.

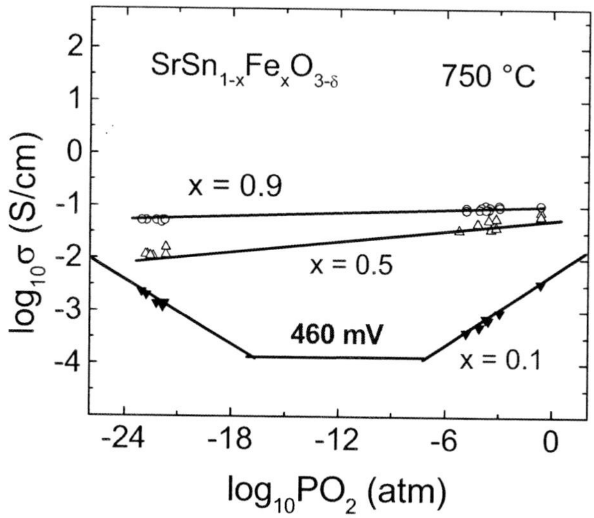

Figure 15. Variation of the total conductivities as function of oxygen partial pressure at 750 °C for $SrSn_{1-x}Fe_xO_{3-\delta}$ (x = 0.1, 0.5 and 0.9). The line connecting the data points is the guide to the eye [89,90].

The open circuit voltage (OCV) at 750 °C for the x = 0.1, 0.15 and 0.2 members of $SrSn_{1-x}Fe_xO_{3-\delta}$ have been found to be 460, 268 and 170 mV, respectively, for air-hydrogen galvanic cells, which proves the existence of predominant ionic conductivity for the oxygen partial pressure range resulting from Nernst's equation. The average transference numbers for oxide ion conduction was found to be in the range 0.15 –0.50 at 600-850 °C (Fig. 16).

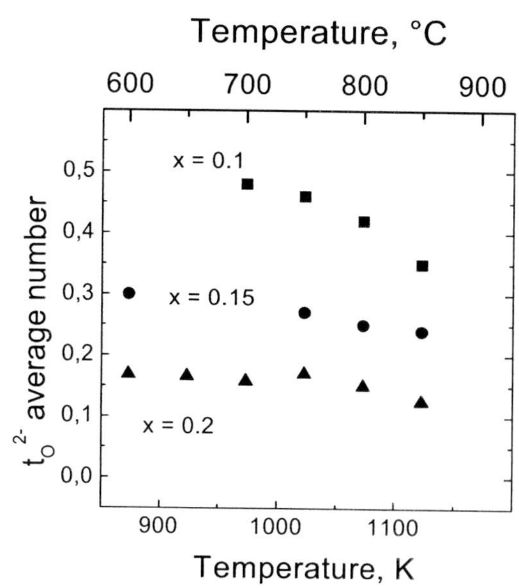

Figure 16. Average oxide ion transference number for $SrSn_{1-x}Fe_xO_{3-\delta}$ (x = 0.1; 0.15 and 0.2) in the temperature range 600-850 °C [90].

The electrical conductivity of the materials containing high Fe ($x \geq 0.5$) concentration does not depend on the oxygen partial pressure over a wide range (Fig. 15). These highly conducting materials are predominantly electronic conductors over the employed oxygen partial pressure regime, which is similar to that of the acceptor doped perovskites. Accordingly, we expect that samples with high Fe contents in $SrSn_{1-x}Fe_xO_3$ may find application as electrodes for SOFCs and sensor applications.

The chemical reaction stability between LSGM and $SrSn_{1-x}Fe_xO_3$ ($x = 0.8, 0.9$) oxides were investigated at 1000 and 1300 °C in air [84b]. Figure 17 shows the powder XRD patterns of reaction products of intimate powder mixtures of LSGM and $SrSn_{1-x}Fe_{0x}O_{3-\delta}$ ($x = 0.8, 0.9$) perovskites. For this purpose, 1.1 weight % LSGM and $SrSn_{1-x}Fe_{0x}O_{3-\delta}$ ($x = 0.8, 0.9$) oxides were mixed and heated at 1000 and 1300 °C for 48 h and cooled to room temperature. We see that LSGM is chemically stable against reaction with $SrSn_{0.1}Fe_{0.9}O_{3-\delta}$ at 1000 °C. We see that the *103* peak (shown as arrow) and other reflections of $LaSrGaO_4$ are absent in products obtained at 1300 °C (e), indicating the formation of new perovskite-like composition.

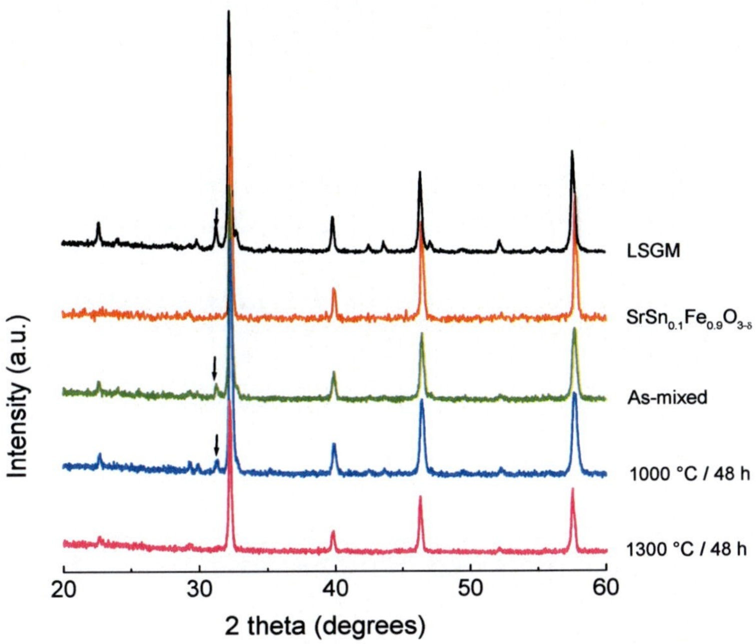

Figure 17. Powder XRD patterns showing the reactivity of intimate mixtures of LSGM and $SrSn_{1-x}Fe_{0x}O_{3-\delta}$ ($x = 0.8, 0.9$) perovskites in air. 1.1 weight % LSGM and $SrSn_{1-x}Fe_{0x}O_{3-\delta}$ ($x = 0.8, 0.9$) oxides were mixed and heated at 1000 and 1300 °C for 48 h and cooled to room temperature. (a) LSGM, (b) $SrSn_{0.1}Fe_{0.9}O_{3-\delta}$, (c) 1:1 weight % sample a and b as-mixed powders, (d) sample c heated at 1000 °C 48 h, (e) sample c heated at 1300 °C 48 h. The arrow mark shows the position of (*103*) reflection of impurity $LaSrGaO_4$ (JCPDS card No.: 83-1004) in LSGM. We see that *103* peak and other reflections is absent in products obtained at 1300 °C (e), indicating the formation of new perovskite-like composition.

4. INVESTIGATIONS ON PARTIAL ELECTRONIC CONDUCTIVITY OF SAMARIUM-DOPED CERIA

Recently, doped CeO_2 has been accepted as potential IT-SOFC electrolyte due its high ionic conductivity compared to YSZ and LSGM in the temperature range 400-600 °C and found to be stable under oxygen partial pressures of 1 - 10^{-19} atm. This range is sufficient to employ ceria as electrolyte in fuel cells with an appropriate selection of the fuel gases [21,24,25,31]. The partial electronic conductivities of RE-doped CeO_2 were estimated at temperatures above 600 °C [94, 95], which is much higher than the actual application temperature. In this section, we report the contribution of the partial electronic conductivity at high oxygen partial pressures (10^{-3}-1 atm) of $Ce_{0.8}Sm_{0.2}O_{1.9}$ (CSO) at 500-600 °C.

Several electrochemical methods are available for the determination of the transference number of the ions and electrons of solid electrolytes, e.g., the DC and EMF method commonly in combination with AC measurements [1,2,8,10,13,17,18,86]. For predominant oxide ion conductors, the total conductivity as a function of oxygen activity will remain constant in AC impedance studies. The EMF method is not applicable to estimate the minority (electrons and holes) charge carriers contribution if that is several orders of magnitude lower than the total conductivity and gives only an average transference number between two different activities of the mobile species. Generally, EMF measurements have to be performed at elevated temperatures in order to obtain equilibria between the sample interfaces and the surrounding gases or solid oxide reference electrodes. The gas phase reference also may result in an incorrect value due to interface reaction products, oxygen gas transfer through the electrolyte, and temperature gradients. In view of these difficulties, we have employed the Hebb-Wagner (HW) [86,94-100] polarization method in combination with AC impedance measurements. The oxide ion blocking Macor® disc was attached to the surface of the CSO pellet using glass-sealing rings (Fig. 18). In addition to the blocking electrode, the surface was covered with high temperature ceramic sealant to avoid possible exchange of oxygen from the surrounding gas.

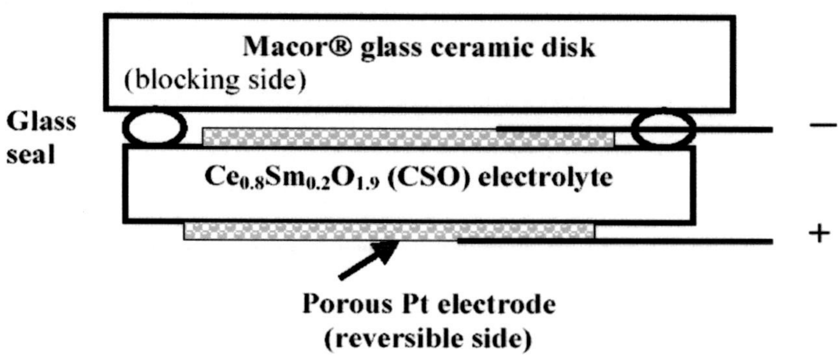

Figure 18. Schematic representation of Hebb-Wagner (HW) polarization measurement. Dense thick Macor® glass attached to the CSO pellet serves as oxide ion blocking layer. Porous Pt-electrodes used as electrodes.

4.1. Total Electrical Conductivity of $Ce_{0.8}Sm_{0.2}O_{1.9}$

$Ce_{0.8}Sm_{0.2}O_{1.9}$ can be prepared by solid-state reaction using appropriate amounts of CeO_2 and Sm_2O_3 at elevated temperature of 1450 °C for 24 h [98]. Powder XRD data confirm the formation of single-phase fluorite type structure with a lattice constant a = 5.433(5) Å (Fig. 19). Figure 20 shows the Arrhenius plots for the bulk ionic conductivity of CSO. A slight change in the slope at about 500 °C may be ascribed to the ordering of oxide ion vacancies [34,35]. A similar change in slope was also reported for other RE-doped CeO_2 (RE = Nd, La, Y, Gd, Yb). The ionic bulk conductivity value obtained in the present work is comparable to that of the reported values in the literature [31].

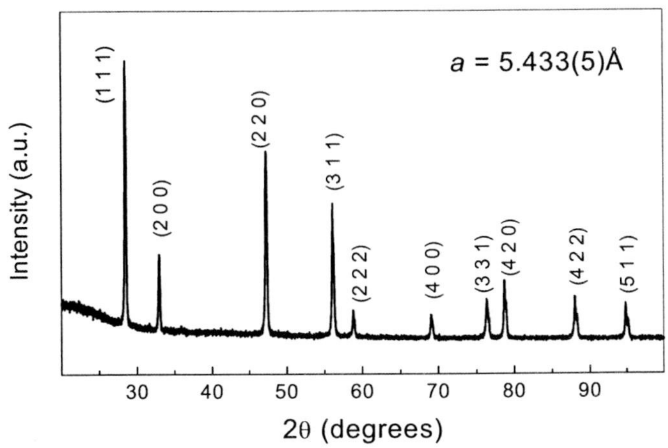

Figure 19. Powder XRD data of CSO prepared by solid-state reaction in air (Seifert 3000 powder X-ray diffractometer (CuK_{α})) [98].

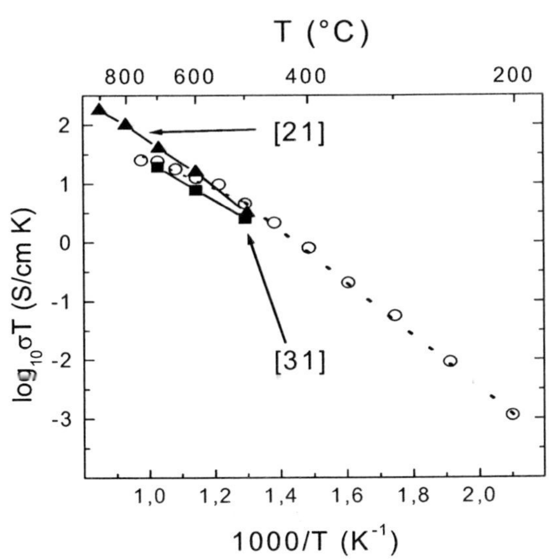

Figure 20. Arrhenius plots for bulk ionic conductivity of Sm-doped CeO_2. For comparison, the data reported in the literature are shown. We see that the present data are similar to those of Steele [31] and we found to be slightly lower at high temperatures compared to the Inaba et. al. [21] values. The activation energy in the low

temperature regime is 0.88 eV (200-550 °C). The broken line passing through the data points is a guide to the eye. A similar change in the slope is also observed for other RE-doped CeO_2 sample and the temperature is found to increase with decrease in RE ionic sizes.

The activation energy, E_a, for the bulk ionic conductivity is found to be 0.88 eV (200-550 °C) and 0.42 eV (500-800 °C). Commonly, it is known for vacancy ordered ionic conductors that E_a consists of two components i.e., $E_a = \Delta H_m + \Delta H_t$, where ΔH_m and ΔH_t represent motional enthalpy and enthalpy associated with trapping of mobile oxide ions, respectively [13]. At high temperature, the oxygen vacancies are disordered and E_a approaches ΔH_m and ΔH_t approaches zero. From our data, the activation energy for the association of defects (ΔH_t) is found to be 0.46 eV.

4.2. Electronic Conductivities of Sm-Doped CeO_2

A voltage lower than the decomposition voltage is applied to the galvanic cell (II):

(-) blocking O_2, Pt|| CSO || Pt, reversible (+) (II)

Since the oxide ion is blocked, only electronic (e^- and $h^\bullet$) currents are passed under steady state conditions. It is essential that no intermediate reaction product is being formed, which might block the electronic current. Also, equilibrium between the gas and the electrolyte at the reversible electrode side is required. We have employed a highly porous Pt-paste. The steady state current has been analysed using the HW equation [94,95]:

$$I = \frac{ART}{LF}\left\{\sigma_h\left(1-\exp\left[\frac{-EF}{RT}\right]\right)+\sigma_e\left(\exp\left[\frac{EF}{RT}\right]-1\right)\right\} \qquad (6)$$

where I, E, σ_e, σ_h, A, L, R, T and F are the electrical current, applied voltage, electron and hole conductivity, electrode-electrolyte interface area, thickness of the sample, gas constant, temperature, and Faraday constant, respectively. The applied voltage will generate a difference in the electronic charge carrier concentration since no electrostatic potential gradient will be generated. This is the case for a predominante ionic conductor employing a mobile ion-blocking electrode where only electronic charge carrier will contribute to the steady state current. The first term in the eqn. 6 approaches a plateau caused by the hole conductivity and the second term causes an exponential increase in current due to the excess electronic conductivity. At very low voltage, $E \ll RT/F$,

$$\exp\left[\frac{\pm EF}{RT}\right] = 1 \pm \frac{EF}{RT} \qquad (7)$$

Accordingly, eqn. 6 becomes

$$I = \frac{A}{L}\sigma_h E \qquad (8)$$

This is Ohm's law, which can be employed to determine the excess electronic (hole) conductivity. The chemical diffusion coefficient could be determined using the delay in the electrical current response upon a variation of the applied voltage with relationship:

$$\tau = \frac{L^2}{\widetilde{D}} \qquad (9)$$

where τ, L and $\widetilde{D}$ are the equilibriation time, the length of the sample and the chemical diffusion coefficient, respectively. Since the rate determining process is the diffusion of the electrons, the chemical diffusion corresponds to the electron diffusion coefficient of CSO. The electron diffusion coefficient values obtained using eqn. (9) is ~ 2 x 10^{-6} cm^2/sec at 500-600 °C using 24 h as equilibration time. This value is comparable with the diffusivity of the oxygen reported for CGO (~10^{-6}-10^{-5} cm^2/sec at 500-970 °C) [21].

The excess electronic conductivity is found to be about three orders of magnitude lower than the total conductivity in air (Fig. 21). Using the diffusion coefficient of 1 x 10^{-6} cm^2/s, the concentration of the electronic defect (hole) is found to be 1.2 x 10^{17}, 4.3 x 10^{17}, 1.04 x 10^{18} cm^{-3} at 500, 550 and 600 °C, respectively, while the oxide ion vacancy concentration (2.5 x 10^{21} cm^{-3}) is 3-4 orders of magnitude higher than that of the holes. This value is comparable with the concentration of electrons and holes in YSZ [99]. Since the diffusion coefficients are nearly of the same order of magnitude, the low conductivity may be attributed to the lower concentrations of holes compared to the oxide ion vacancies in CSO. In Fig. 21, we show the variation of hole conductivity as a function of temperature in oxygen, air and 1000 ppm oxygen gas. The activation energy for the hole conduction is found to be 1.2-1.6 eV, which is comparable to those of other RE-doped CeO_2 [100-104] (Table 4).

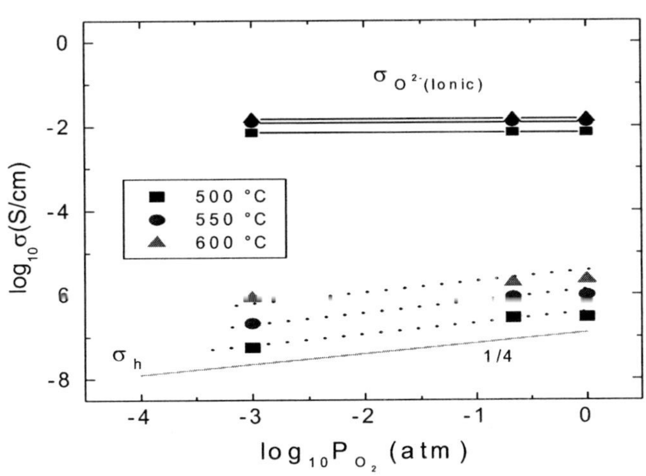

Figure 21. Top: Comparison of electronic (hole) conductivities of CSO with bulk ionic conductivity in air. Bottom: Dependencies of excess hole conductivity (open symbols) of CSO at 500, 550 and 600 °C as a function of oxygen activity. The lines passing through the experimental point is a guide to the eye and slope of 1/4 is shown for comparison.

Table 4. Comparison of the electronic (hole) partial conductivities of CSO with hole conductivity of other predominate ionic conductors

Compound	σ, T (S/cm), °C	Oxygen partial pressure (atm)	$E_{a,h}$ (eV)	Reference
$Ce_{0.8}Sm_{0.2}O_{1.9}$ (CSO)	9.68 x 10^{-7}, 550	1	1.24	98
	2.07 x 10^{-7}, 550	1 x 10^{-3}	1.61	98
$Ce_{0.8}Gd_{0.2}O_{1.9}$ (CGO)	4.4 x 10^{-6}, 600	1	1.07	95
	~ 10^{-6}, 600	1 x 10^{-3}	1.28	95
LSGM	3.44 x 10^{-4},800	0.21	1.12	101
	1 x 10^{-4}, 600	1	1.06	102
	2.5 x 10^{-5}, 600	7 x 10^{-3}	1.07	102
YSZ	1 x 10^{-6}, 650	0.21	1.4	103
$YO_{1.5}$–CeO_2	1.7 x 10^{-3}, 1000	1 x 10^{-1}	1.24	104

4.3. Limiting Current Amperometric Oxygen Sensor Study

A schematic representation of the experimental set up used for the amperometric (limiting current, I_L) oxygen sensor is shown in Fig. 22. A diffusion barrier (0.16 mm in diameter, 1 mm in height) was attached to one surface of the pellet using a glass ring. The measurements were performed in the range 2-1000 ppm oxygen gas at 550 °C. The required gas concentration was obtained by diluting commercially available 1000 ppm and 2 ppm O_2 in Ar gas using mass flow controllers. The I_L is determined by the oxygen gas diffusion through the diffusion barrier. The magnitude of the limiting current depends on the oxygen partial pressure, which can be calculated by using the modified Fick's law [105]:

$$I_L = -\frac{4FDAP}{RTL}\ln\left(1-\frac{P_{O_2}}{P}\right) \tag{10}$$

where D, A, L, P_{O2}, and P are the oxygen diffusion coefficient in the gas, the area and the length of the diffusion hole, the oxygen partial pressure and the total pressure of the gases, respectively.

A fast response of the current to the voltage was observed first and followed by a small decrease. The I-V characteristic curves of the limiting-current (I_L) sensor are shown in Fig. 23. A sharp increase in the electronic current was observed above 800 mV for all oxygen concentrations used. This may be due to the decomposition or the electronic leakage of the CSO electrolytes. A similar behaviour was observed in the HW measurements due to the electronic current in CSO. The response of the CSO sensor to varying concentrations of oxygen at 550 °C with an applied voltage of 400 mV is shown in Fig. 24. We see that the current obtained during increasing and decreasing oxygen concentrations follows the same line, indicating a good reproducibility of the sensor current. We see a fast response first and a steady-state within 30 min and a steady state current after more than 3 days. The measured

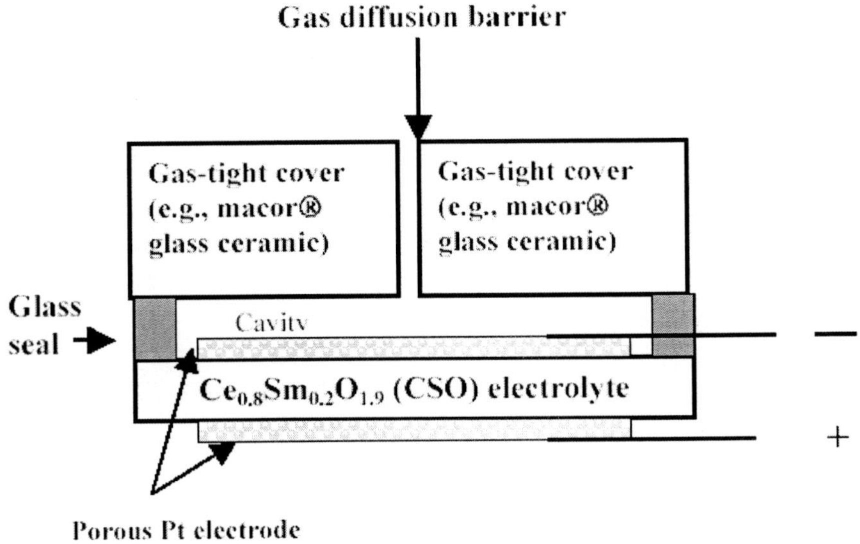

Figure 22. Schematic representation used for limiting current sensor study. A gas diffusion barrier of about 0.16 mm in diameter and 1 mm in height was attached to CSO electrolyte using glass ring seals.

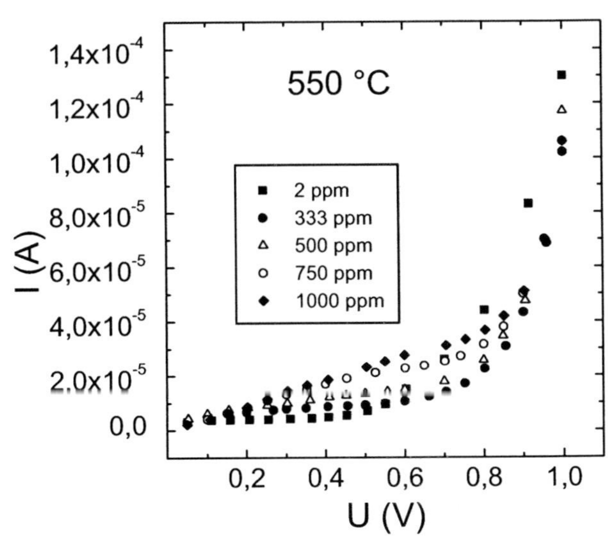

Figure 23. A steady-state current of oxygen sensor as a function of applied voltage at a few oxygen partial pressures at 550 °C [98].

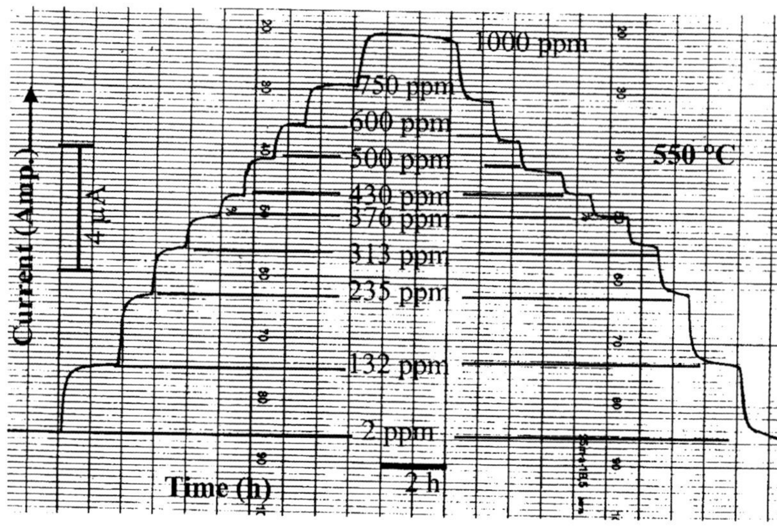

Figure 24. A typical response of CSO oxygen sensors to the oxygen concentration at 550 °C at applied voltage of 400 mV [98].

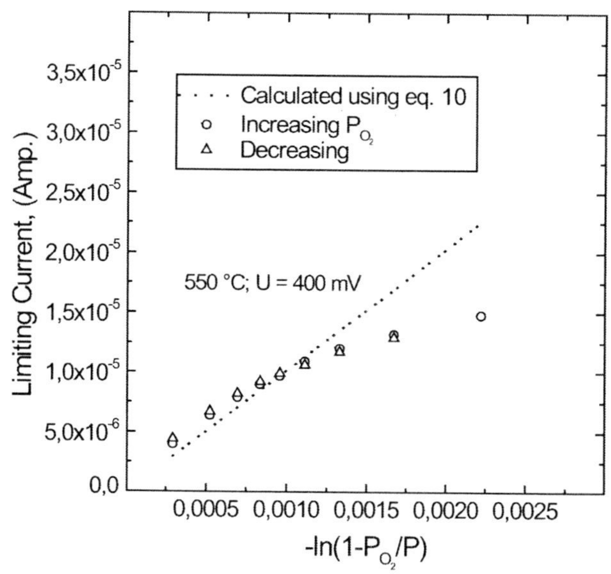

Figure 25. Steady state current as a function of oxygen partial pressures at 550 °C. The broken line represents the calculated limiting current using eqn. 10. at 550 °C [98].

and calculated I_L using eqn. (10) are shown together with the experimental data in Fig. 25. A linear relationship was observed for 100-500 ppm oxygen gas concentration. At high oxygen concentration, we see that the measured current is lower than the calculated ones.

5. STUDIES OF SCHEELITE-TYPE (ABO_4) STRUCTURED MATERIALS

Scheelite-type ABO_4 (A = Ca, Sr, Ba, Pb; B = Mo, W) materials have been prepared at room temperature in aqueous media without any subsequent annealing. 0.06 mol of the material were precipitated by dissolving high purity (> 99.5 %) alkaline earth and lead chloride in 50 ml of distilled water and adding this solution to the solution of appropriate amounts of sodium molybdate or tungstate under continuous stirring. A white precipitate was formed spontaneously, which was filtered and washed with a large amount of warm distilled water. The products were dried over night in an oven with air atmosphere at 100 °C. The yield was always better than 95 wt-% (except in the case of the Ca- members) [38].

Figure 26 shows powder XRD data of the as-precipitated products and confirms the formation of scheelite-type structure compounds, which correspond to crystalline ABO_4 scheelite-type phases [38,39]. The structure is related to common alkali metal molybdates and tungstates with the general formula M_2BO_4 (M = Na; B = Mo, W), which contain discrete tetrahedral BO_4^{2-} units in crystals and in aqueous solution [106]. Accordingly, several molybdates and tungstates of other metals can be prepared by metathetical reactions of the type:

$$\text{aq. } M_2^{I}BO_4 + \text{aq. } ACl_2 \longrightarrow ABO_4 \text{ (ppt)} + \text{aq. } 2MCl \tag{11}$$

Indeed, the reaction occurs already at room temperature in an aqueous medium to from scheelite-type ABO_4 (Fig. 26), in which tetrahedral BO_4^{2-} units exist. These are the primary units for this structure. In an ideal scheelite structure, both A and B cations are ordered with an FCC arrangement. Due to the difference in cation sizes, the anion lattice is distorted. Larger A (Sr,Ba,Pb,La) cations show 8-fold coordination and smaller B (Mo, W, V) cations show 4-fold coordination [107]. The readily occurring reaction (11) at room temperature suggests that most likely a topotactic metathesis occurs, since the tetrahedral BO_4^{2-} units are retained during the chemical transformation. Table 5 gives the lattice parameters of all the investigated scheelites. The reaction time is completed within a minute, which is advantageous over other isostructural preparation techniques of compounds, e.g., with the formula $BiVO_4$ by solid state reaction and by precipitation by the addition of the basic solution of Na_2CO_3, $NaHCO_3$ or NH_4OH to an aqueous nitrate solution mixture of $Bi(NO_3)_3$ and $Na_3VO_4.7H_2O$ and stirring for 4 - 46 h before filtering and drying.

The particle sizes were determined with a measured full width half maxima (FWHM) value of the (100 %) *112* diffraction peak using the Scherrer formula

$$t = \frac{0.9\,\lambda}{B\cos\theta} \tag{12}$$

where t is size of the particle , λ is the X-ray wavelength (0.15418 nm), θ is the Bragg angle and B is the FWHM. The value of B is calculated by using the formula

$$B^2 = B_m^2 - B_S^2 \tag{13}$$

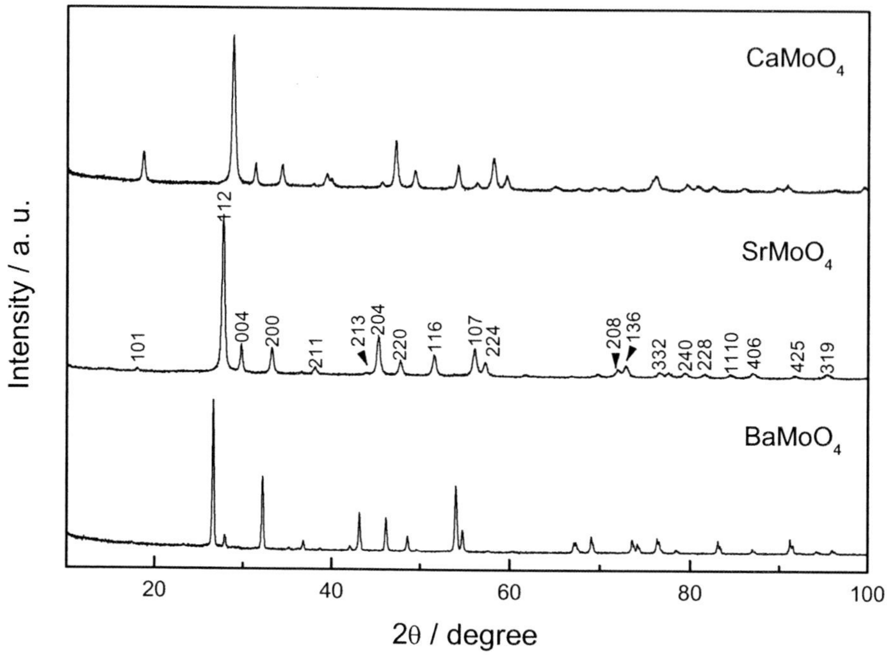

Figure 26. Powder XRD patterns of $AMoO_4$ (A = Ca, Sr, Ba) prepared by metathesis reaction using aqueous alkaline earth chloride and sodium molybdate at room temperature [38].

Table 5. Composition, particle size and comparison of tetragonal lattice constants (Å) of simple molybdates and tungsates with the general formula ABO_4 (A = Ca, Sr, Ba, Pb; B = Mo, W) prepared at room temperature by metathesis reactions [38]

Compound	Particle size[a] (nm)	Lattice constant		Lattice constants reported in JCPDS		
		a	c	a	c	JCPDS No:
$CaMoO_4$	21	5.221(6)	11.434(4)	5.2235(1)	11.4298(4)	85-1267
$SrMoO_4$	24	5.390(6)	12.012(3)	5.3944	12.020	08-0482
$BaMoO_4$	51	5.562(2)	12.818(8)	5.5802	12.821	29-0193
$CaWO_4$	12	5.238(2)	11.364(5)	5.2425(5)	11.3715(5)	72-1624
$SrWO_4$	17	5.406(3)	11.92(1)	5.4168	11.951	08-0490
$BaWO_4$	51	5.604(9)	12.697(5)	5.6123(2)	12.7059(9)	43-0646
$PbWO_4$	67	5.445(2)	12.01(8)	5.44503	12.0495	85-1857

[a] Calculated from XRD data.

where B_m is the measured FWHM and B_s is the FWHM of a standard Si sample. The calculated particle sizes are in the range 12-67 nm and are also listed in Table 4 along with the lattice parameters.

TGA (Netzsch Thermal Analyzer, model STA 409), data show that the prepared scheelites do not show significant weight losses due to the presence of water and are structurally stable up to the highest investigated temperature of 1000 °C, which is consistent with Fourier transform infrared spectrometry (FTIR) (Nicolet model, using KBr) results, which show no band corresponding to the M-O-H bond.

SEM (PHILIPS SEM XL 30) studies of $AMoO_4$ precipitated at room temperature were carried out and photographs are presented in Fig. 27 together with that of $CaWO_4$. The crystalline particles are spheres with diameters less than 10 μm for the Ca- and Sr-members and needle like structures in the case of the Ba-compound. This value is higher than that derived from powder XRD patterns suggesting that the spheres may contains an aggregated of small nanocrystallites.

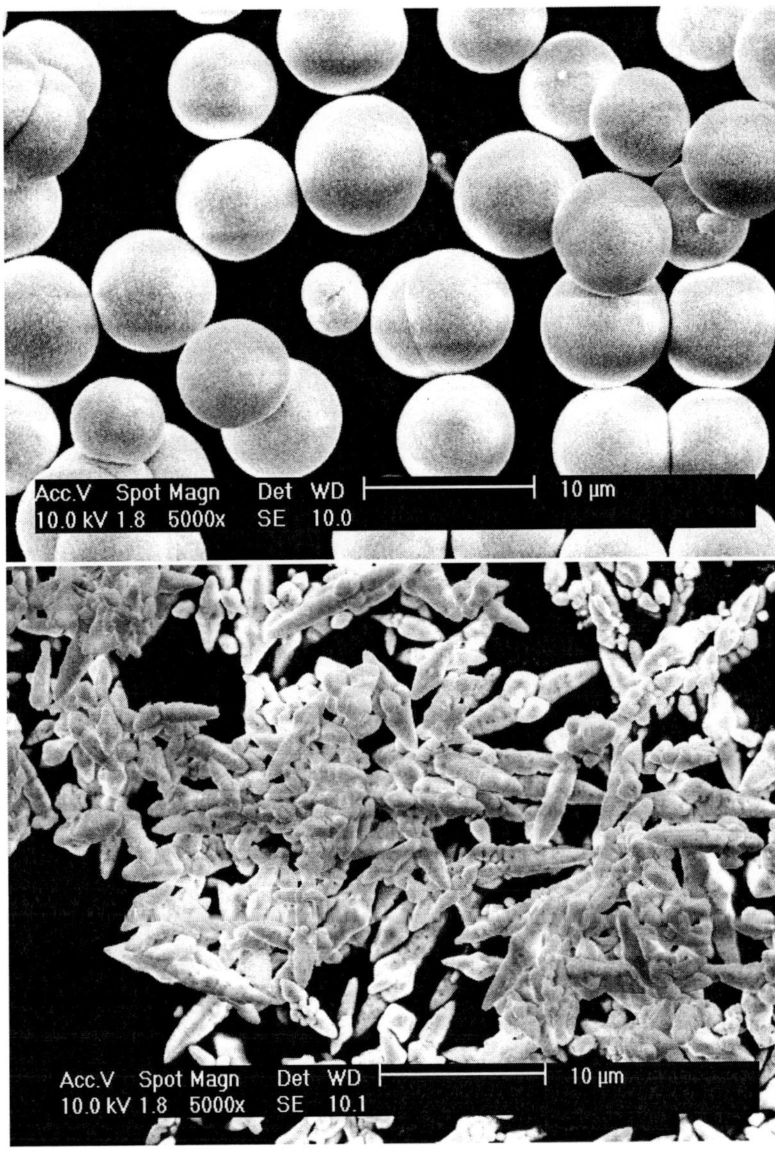

Figure 27. SEM photographs of as-prepared (top) $CaMoO_4$ and (bottom) $BaMoO_4$ powders [38].

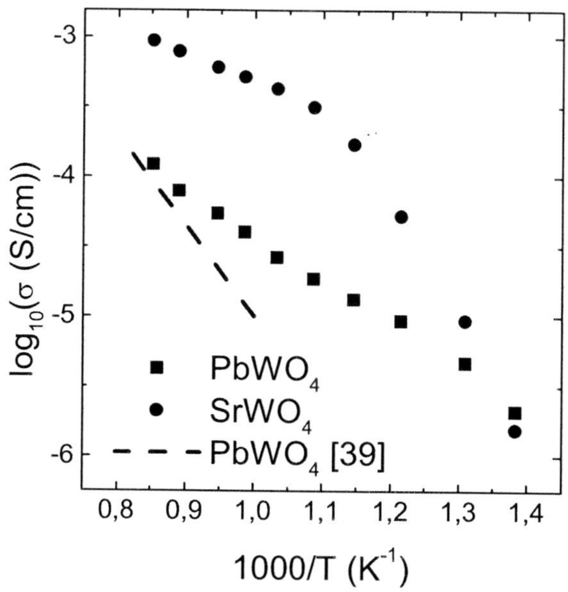

Figure 28. Arrhenius plots of the electrical conductivities of $SrWO_4$ and $PbWO_4$. For comparison, the electrical conductivity data for solid state synthesized $PbWO_4$ (---) is also shown [ref. 39]. The data obtained from first cycle cooling and subsequent heating runs follow the same line.

The Arrhenius plots for the equilibrium electrical conductivity are shown in Fig. 28. The data obtained from first cycle cooling and subsequent heating runs follow the same line with activation energies of 0.40 and 0.62 eV for $SrWO_4$ (650-950 °C) and $PbWO_4$ (500-900 °C), respectively. $PbWO_4$ shows no transition in the conductivity up to 900 °C. The conductivity data of $PbWO_4$ are the same as those of solid state synthesized samples at high temperatures [24], and are lower than that of $SrWO_4$ at higher temperatures (Fig. 28). The difference in the activation energy of the solid-state and the present room temperature synthesized $PbWO_4$ may be due to the difference in the oxygen activity. The activation energy is similar to those of other perovskite type oxide ion conducting electrolytes [30]. Oxygen partial pressure dependence of the conductivity at 735 °C shows p-type electronic conductivity with a slope of 1/6 over the investigated partial pressure range (Fig. 29).

The following galvanic cell (III):

$$Ni,NiO||SrWO_4||CuO,Cu_2O \qquad \text{(III)}$$

has been constructed to determine the transference number for oxide ion conduction. The EMF of the galvanic cell is shown in Fig. 30 for various temperatures together with the average transference number of the oxide ion for the given oxygen partial pressure range (Fig. 30b) which is obtained by using eqn. (3). The value is found to be 0.86-0.95 over the temperature range 600-850 °C.

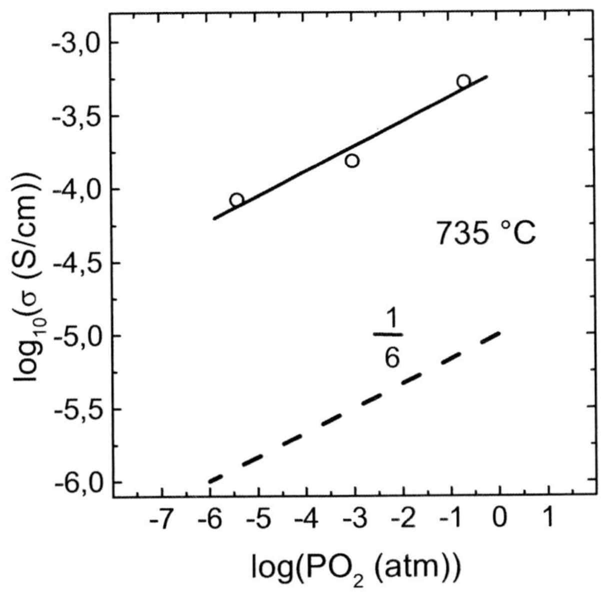

Figure 29. Oxygen partial pressure dependent electrical conductivity of $SrWO_4$ at 735 °C showing p-type electronic conduction with a slope of 1/6. For comparison, the slope 1/6 is shown [38].

6. Conclusions and Outlook

Perovskite-type (ABO_3) compounds form one of the most important and fascinating classes of materials with interesting functional properties among the various transition metal oxides. The perovskite structure is very tolerant against changes in the ionic radii of both the A and B-cations and can also accommodate large concentrations of cation vacancies, particularly at the A-site, as well as oxygen vacancies. The oxide ion conductivity is dependent on the presence of oxide ion vacancies and the nature of A and B-site cations. For example, among the $A^{III}B^{III}O_3$ perovskite-type compounds, Sr+Mg-doped $LaGaO_3$ perovskite exhibits high oxide ion conductivity over a wide range of oxygen partial pressures compared to that of the Al, Sc and In analogues. The x = 0.1 and y = 0.2 member of the chemical composition series $La_{1-x}Sr_xGa_{1-y}Mg_yO_3$ (LSGM), shows an ionic conductivity of ~ 0.1 S/cm at 800 °C with an activation energy of 1.07 eV. Attempts have been made to improve the ionic conductivity by partial or complete substitution of La or Ga or Mg by iso- or heterovalent (or aliovalent) metal ions. LSGM remains unique with regard to the chemical composition-structure-electrical conductivity relationships. No major improvement has been made so far.

Rare earth (RE)-doped CeO_2 appears to belong to another important family of materials for IT-SOFC developments due its high ionic conductivity at lower temperature compared to that of presently employed solid electrolyte YSZ in the SOFCs. Further optimization of the electrical properties by chemical substitutions, synthesis conditions, heat treatments, control the oxygen activities and operation temperature are crucial issues for future developments of long-term stable fuel cells. Further, attention must be paid to the layered or tunnel structure possessing stable cations, in which large concentration of oxygen vacancies may be introduced by chemical substitutions.

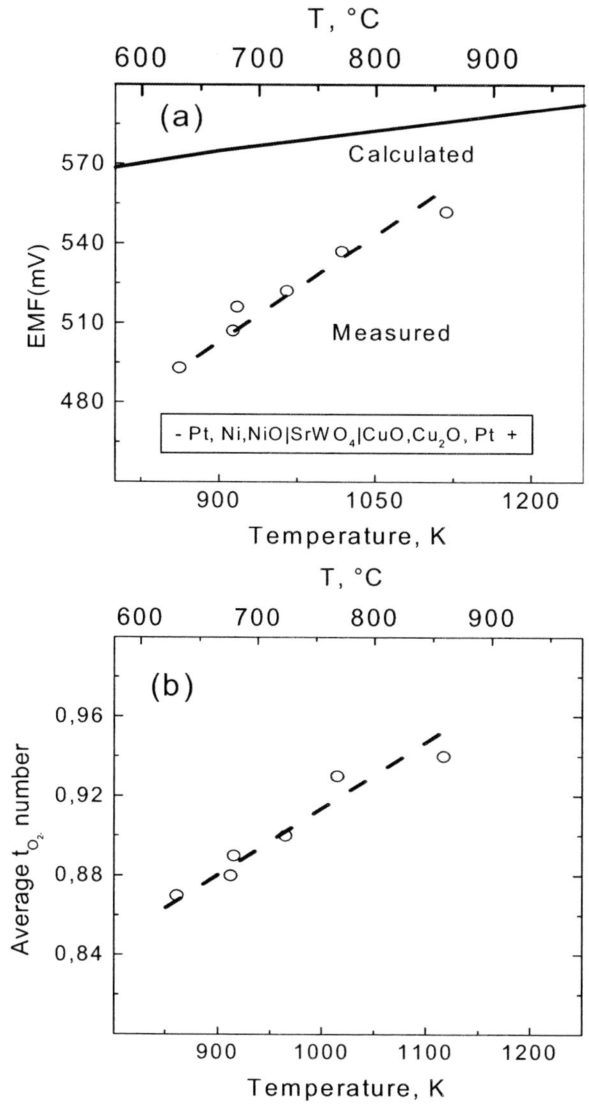

Figure 30. (a) Equilibrium OCV of the galvanic cell and (b) average transference number in the temperature range 650-850 °C. The line connecting the data points is a guide to the eye [38].

In general, the practically applicable solid oxide ion conductors have to be exposed to two different chemical potentials (anode and cathode) of oxygen in all practical application. Hence, solid electrolytes have to be chemically and electrochemically stable for a long period of time without showing interfacial chemical reactions. Further investigations, both from thermodynamic and kinetic points of view are required, especially since all voltages are generated only over very narrow regimes at the electrolyte/electrode interfaces.

ACKNOWLEDGEMENTS

Dr. V. Thangadurai would like to thank the German Academic Exchange Service (DAAD) for financial support in the frame of Guest Professorhip and the Alexander von Humboldt (AvH) Foundation for support during the initial stages of the present work in the form of Research Fellowship.

REFERENCES

[1] Rickert, H. (1978). Solid State Ionic Conductors: Principles and Applications. *Angew. Chem. Int. Ed. Engl.*, *17*, 37-46.

[2] Hagenmuller, P., & Van Gool, W., (Ed.), (1978). Solid Electrolytes General Principles, Characterization, Materials, Applications. New York: Academic Press.

[3] Chandra, S. (1981). Superionic Solids: Principles and Applications, Amsterdam: North Holland.

[4] West, A. R. (1989). Solid Electrolytes. *Ber. Bunsenges. Phys. Chem.*, *93*, 1235-1241.

[5] Takahashi, T. (1995). Solid State Ionics: Recent Trends and Future Expectations. *Bulletin of Electrochemistry*, *11*, 1-33.

[6] Greenblatt, M. (1994). Ionic Conductors. In R. B. King (Ed.), *Encyclopaedia of Inorganic Chemistry* (pp. 1584-1602), Chichester: John Wiley.

[7] Birke, P., & Weppner, W. (1999). Solid Electrolytes. In J. O. Besenhard (Ed.), Handbook of Battery Materials (pp. 525-552). Weinheim: Wiley-VCH.

[8] Subbarao, E. C., (Ed.), (1980). Solid Electrolytes and Their Applications, New York: Plenum Press.

[9] Takahashi, T., & Kozawa, A. (Ed.), (1980). Applications of Solid Electrolytes. Washington: JEC Press,

[10] Rickert, H. (1982). Electrochemistry of Solids: An Introduction, Berlin: Springer-Verlag.

[11] Takahashi, T. (Ed.), (1989). High Conductivity Solid Ionic Conductors, Recent Trends and Applications. Singapore: World Scientific.

[12] Julien, C., & Nazri, G. A. (1994). Solid State Batteries: materials Design and Optimization. Boston: Kluwer Academic Publishers.

[13] Bruce, P. G., (Ed.), (1995). Solid State Electrochemistry, Cambridge: Cambridge University Press.

[14] (a) Gellings, P. J., & Bouwmeester, H. J. M. (Ed.,). (1997). The CRC Handbook of Solid State Electrochemistry. Boca Raton: CRC Press. (b) Thangadurai, V., & Weppner, W. (2003). Materials Aspects of Solid State Ionic Devices: Solid Oxide Fuel Cells and Electrochromism, In Recent. Res. Devel. Solid State Ionics, 1, pp.131-156, Trivandrum, Kerala: Transworld Research Network..

[15] Brown, J. T. (1989). Solid Oxide Fuel Cells. In T. Takahashi (Ed.), High Conductivity Solid Ionic Conductors Recent Trends and Application (pp. 630-663), Singapore: World Scientific.

[16] (a) Nernst, W. (1897), Verfahren zur Erzeugung von elektrischem Glühlicht, German Patent. No. 104872. (b) Nernst, W. (1899). Über die elektrolytische Leitung fester Körper bei sehr hohen Temperaturen. *Z. Elektrochem.*, 6, 41-43.

[17] Kiukkola, K., & Wagner, C. (1957). Galvanic Cells for the Determination of the Standard Molar Free Energy of Formation of metal Halides, Oxides, and Sulfides at Elevated Temperatures. *J. Electrochem. Soc.*, *104*, 308-316.

[18] Kiukkola, K., & Wagner, C. (1957). Measurements on Galvanic Cells Involving Solid Electrolytes. *J. Electrochem. Soc.*, *104*, 379-386.

[19] Etsell, T. H., & Flengas, S. N. (1970). The Electrical Properties of Solid oxide Electrolytes. *Chem. Rev.*, *70*, 339-376.

[20] Steele, B. C. H. (1989). Oxide Ion Conductors. In T. Takahashi (Ed.), High Conductivity Solid Ionic Conductors Recent Trends and Application (pp. 402-446), Singapore: World Scientific.

[21] Inaba, H., & Tagawa, H. (1996). Ceria Based Solid Electrolytes. *Solid State Ionics*, *83*, 1-16.

[22] Shuk, P., Wiemhöfer, H. D., Guth, U., Göpel, W., & Greenblatt, M. (1996). Oxide Ion Conducting Solid Electrolytes Based on Bi_2O_3. *Solid State Ionics*, *89*, 179-196.

[23] Kharton, V.V., Naumovich, E. N., & Vecher, A. A. (1999). Research on the Electrohemistry of Oxide Ion Conductors in the Former Soviet Union. I. ZrO_2-based Ceramic Materials, *J. Solid State Electrochem.*, *3*, 61-81.

[24] Kharton, V.V., Yaremchenko, A. A., Naumovich, E. N., & Marques, F. M. B. (2000). Research on the Electrohemistry of Oxide Ion Conductors in the Former Soviet Union. III. HfO_2, CeO_2 and ThO_2-Based Oxides. *J. Solid State Electrochem.*, *4*, 243-266.

[25] Boivin, J. C., & Mairesse, G. (1998). Recent Materials Developments in Fast Oxide Ion Conductors. *Chem. Mater.*, *10*, 2870-2888.

[26] Kharton, V.V., Yaremchenko, A. A., & Naumovich, E. N. (1999). Research on the Electrohemistry of Oxide Ion Conductors in the Former Soviet Union. II. Perovskite-Related Oxides. *J. Solid State Electrochem.*, *3*, 303-326.

[27] Kendall, K. R., Navas, C., Thomas, J. K., & zur Loye, H. C. (1996). Recent Developments in Oxide Ion Conductors: Aurivillius Phases. *Chem. Mater.*, *8*, 642-649.

[28] Kendall, K. R., Navas, C., Thomas, J. K., & zur Loye, H. C. (1995). Recent Developments in Perovskite-Based Oxide Ion Conductors. *Solid State Ionics, 82*, 215-223.

[29] Hayashi, H., Inaba, H., Matsuyama, M., Lan, N. G., Dokiya, M., & Tagawa, H. (1999). Structural Consideration on the Ionic Conductivity of Perovskite-Type Oxides. *Solid State Ionics*, *122*, 1-15.

[30] Kudo, T., & Fueki, K. (1990). Solid State Ionics. VCH: Basel.

[31] Steele, B. C. H. (2000). Appraisal of $Ce_{1-x}Gd_yO_{2-y/2}$ Electrolytes for IT-SOFC Operation at 500 °C. *Solid State Ionics*, *129*, 95-110.

[32] Kramer, S. A., & Tuller, H. L. (1995). A Novel Titanate-Based Oxygen Ion Conductor: $Gd_2Ti_2O_7$. *Solid State Ionics, 82*, 15-23.

[33] Takahashi, T., & Iwahara, H. (1971). Ionic Conduction in Perovskite-Type Oxide Solid Solution and its Application to the Solid Electrolyte Fuel Cell. *Energy Conversion, 11*, 105-111.

[34] Ishihara, T., Matsuda, H., & Takita, Y. (1994). Doped $LaGaO_3$ Perovskite Type Oxide as a New Oxide Ionic Conductor. *J. Am. Chem. Soc.*, *116*, 3801-3803.

[35] Feng, M., & Goodenough, J. B. (1994). A Superior Oxide-Ion Electrolyte. *Eur. J. Solid State Inorg. Chem.*, *31*, 663-672.

[36] Goodenough, J. B., Ruiz-Diaz, J. E., & Zhen, Y. S. (1990). Oxide-Ion Conduction in $Ba_2In_2O_5$ and $Ba_3In_2MO_8$ (M = Ce, Hf, or Zr). *Solid State Ionics*, *44*, 21-31.

[37] Yao, T., & Uchimoto, Y., Kinuhata, M., Inagaki, T., & Yoshida, H. (2000). Crystal Structure of Ga-doped $Ba_2In_2O_5$ and its Oxide Ion Conductivity. *Solid State Ionics*, *132*, 189-198.

[38] Thangadurai, V., Knittlmayer, C., & Weppner, W. (2004). Metathetic Room Temperature Preparation and Characterization of Scheelite-Type ABO_4 (A = Ca, Sr, Ba, Pb, B = Mo, W) Powders. *Mater. Sci. Eng. B. 106*, 228-233.

[39] Esaka , T. (2000). Ionic Conduction in substituted Scheelite-Type Oxide. *Solid State Ionics*, *136-137*, 1-9.

[40] Abraham, F., Debreuille-Gresse, M. F., Mairesse, G., & Nowogrocki, (1988). Phase-Transitions and Ionic Conductivity in $Bi_4V_2O_{11}$ An Oxide With a Layered Structure. *Solid State Ionics 28-30*, 529-532.

[41] Goodenough, J.B., Manthiram, A., Paranthaman, P., Zhen, Y. S. (1992). Oxide Ion Electrolytes. *Mater. Sci. Eng B12*, 357-364.

[42] Goodenough, J.B., Manthiram, A., Paranthaman, P., Zhen, Y. S. (1992). Fast Oxide-Ion Conduction in Intergrowth Structures. *Solid State Ionics 52*, 105-109.

[43] Kendall, K. R., Thomas, J. K., & zur Loye, H. C. 81995). Synthesis and Ionic Conductivity of a New Series of Modified Aurivillius Phases. *Chem. Mater.*, *7*, 50-57.

[44] Nakayama, S., & Sakamoto, M. (1998). Electrical Properties of New Type High Oxide Ionic Conductor $RE_{10}Si_6O_{27}$ (RE = La, Pr, Nd, Sm, Gd, Dy). *J. Eur. Ceram. Soc.*, *18*, 1413-1418.

[45] Nakayama, S., Kageyama, T., Aono, H., & Sadaoko, Y. (1995). Ionic Conductivity of Lanthanoid Silicates, $Ln_{10}(SiO_4)_6O_3$ (Ln = La, Nd, Sm, Gd, Y, Ho, Er and Yb). *J. Mater. Chem.*, *5*, 1801-1805.

[46] Nakayama, S., Aono, H., & Sadaoko, Y. (1995). Ionic Conductivity of $Ln_{10}(SiO_4)_6O_3$ (Ln = La, Nd, Sm, Gd, and Dy). *Chem. Letter.*, *24*, 431-442.

[47] Martin-Sedeno, M. C., Losilla, E. R., Leon-Reina, L., Bruque, S., Marrero-Lopez, D., Nunez, P., & Aranada, M. A. G. (2004). Enhancement of Oxide Ion Conductivity in Cuspidine-Type Materials. *Chem. Mater.*, *16*, 4960-4968.

[48] Wells, A. F. (1984). Structural Inorganic Chemistry, 5th Edition, Oxford: Clarendon Press.

[49] Ando, M., Enoki, M., Nishiguchi, H., Ishihara, T., & Takita, Y. (2004). Oxide Ion Conductivity and Chemical Stability of Lanthanum Fluorides Doped with Oxygen, $La(Sr,Na)F_{3-2x}O_x$. *Chem. Mater.*, *16*, 4109-4115.

[50] (a) Ishihara, T., Arikawa, H., Akbay, T., Nishiguchi, H., & Takita, Y. (2001). Nonstoichiometric $La_{2-x}GeO_{5-\delta}$ Monoclinic Oxide as a new Fast Oxide Ion Conductor. *J. Am. Chem. Soc.*, *123*, 203-209. (b) Lacorre, P., Goutenoire, F., Bohnke, O., Retoux, R., & Laligant. Y. (2000). Designing fast Oxide-Ion Conductors Based on $La_2Mo_2O_9$. *Nature, 404*, 856-858.

[51] Minh, N. Q. (1993). Ceramic Fuel Cells. *J. Am. Ceram. Soc.*, *76*, 563-588.

[52] Singhal, S. C. (2000). Advances in Solid Oxide Fuel Cell Technology. *Solid State Ionics*, *135*, 305-313.

[53] Singhal, S. C. (2000). Science and Technology of Solid Oxide Fuel Cells. *MRS Bulletin*, *25*, 16-21.

[54] Singhal, S. C. (2002). Solid Oxide Fuel Cells for Stationary, Mobile, and Military Applications. *Solid State Ionics*, *152-153*, 405-410.

[55] Tietz, F., Buchkremer, H. P., & Stöver, D. (2002). Component Manufacturing for Solid Oxide Fuel Cells. *Solid State Ionics*, *152-153*, 373-381.

[55] Dokiya, M. (2002). SOFC System and Technology. *Solid State Ionics*, *152-153*, 383-392.

[56] Perry, M. L., & Fuller, T. F. (2002). A Historical Perspective of Fuel Cell Technology in the 20th Century. *J. Electrochem. Soc.*, *149*, S59-S67.

[57] Williams, M. C., Strakey, J. P., & Singhal, S. C. (2004). U. S. Distributed Generation Fuel Cell Program. *J. Powder Sources*, *131*, 79-85.

[58] Tu, H., & Stimming, U. (2004). Advances, Aging Mechanisms and Lifetime in Solid Oxide Fuel Cells. *J. Powder Sources*, *127*, 282-293.

[59] Weber, A., & Ivers-Tiffee, E. (2004). Materials and Concepts for Solid Oxide Fuel Cells (SOFCs) in Stationary and Mobile Applications. *J. Powder Sources*, *127*, 373-283.

[60] Atkinson, A., Barnett, S., Gorte, R. J., Irvine, J. T. S., Mcevoy, A. J., Mogensen, M., Singhal, S. C., & Vohs, J. (2004). Advanced Anodes for High-Temperature Fuel Cells. *Nature. Mater.*,*3*, 17-27.

[61] Föger, K., & Love, J. G. (2004). Fifteen Years of SOFC Developments in Australia. *Solid State Ionics*, *174*, 119-126.

[62] (a) Kharton, V. V., Marques, F. M. B., & Atkinson, A. (2004). Transport Properties of Solid Oxide Electrolyte Ceramics: A Brief Review. *Solid State Ionics*, *174*, 135-149. (b) Wincewicz, K. C., & Cooper, J. S. (2005). Taxonomies of SOFC Material and Manufacturing Alternatives. *J. Power Sources, 140*, 280-296.

[63] Yokokawa, H., Sakai, N., Kawada, T., & Dokiya, M. (1991) Thermodynamic Analysis on Reaction Profiles between $LaMO_3$ (M=Ni, Co, Mn) and ZrO_2. *J. Electrochem. Soc.*, *138*, 2719- 2726.

[64] Kawada, T., Sakai, N., Yokokawa, H., & Dokiya, M. (1992). Reaction Between Solid Oxide Fuel Cell Materials. *Solid State Ionics*, *50*, 189-196.

[65] Brugnoni, C., Ducati, U., & Scagliotti, M. (1995). SOFC Cathode / Electrolyte Interface. Part I: Reactivity Between $La_{0.85}Sr_{0.15}MnO_3$ and ZrO_2-Y_2O_3. *Solid State Ionics*, *76*, 177-182.

[66] Tu, H., Takeda, Y., Imanishi, N., & Yamamoto, O. (1997). $Ln_{1-x}Sr_xCoO_3$ (Ln = Sm, Dy) for the Electrode of Solid Oxide Fuel Cells. *Solid State Ionics*, *100*, 283-288.

[67] Mori, M., Abe, T., Itoh, H., Yamamoto, O., Shen, G. Q., Takeda, Y., & Imanishi, N. (1999). Reaction Mechanism Between Lanthanum Maganite and Yttria Doped Cubic Zirconia. *Solid State Ionics*, *123*, 113-119.

[68] Brandon, N. P., Skinner, S., & Steele, B. C. H. (2003). Recent Advances in Materials for Fuel Cells. *Annu. Rev. Mater. Res.*, *33*, 183-213.

[69] Yang, Z., Meinhardt, K. D., & Stevenson, J. W. (2003). Chemical Compatibility of Barium-Calcium-Aluminosilicate-Based Seating Glasses with the Ferritic Stainless Steel Interconnect in SOFCs. *J. Electrochem. Soc.*, *150*, A1095-A1101.

[70] Park, S., Vohs, J. M., & Gorte, R. (2000). Direct Oxidation of Hydrocarbons in a Solid-Oxide Fuel Cell. *Nature*, *404*, 265-266.

[71] Hibino, T., Hashimoto, A., Inoue, T., Tokuno, J., Yoshida, A., & Sano, M. (2000). A Low-Operating-Temperature Solid Oxide Fuel Cell in Hydrocarbon-Air Mixtures. *Science*, *288*, 2031-2033.

[72] Zha, S., Moore, A., Abernathy, H., & Liu, M. (2004). GDC-Based Low-Temperature SOFCs Powered by Hydrocarbon Fuels. *J. Electrochem. Soc.*, *151*, A1128-A1133.

[73] Ishihara, T., Matsuda, H., & Takita, Y. (1994). Oxide Ion Conductivity in Doped $NdAlO_3$ Perovskite-Type Oxide. *J. Electrochem. Soc.*, *141*, 3444-3449.

[74] Hashimoto, S., Kishimoto, H., & Iwahara, H. (2001). Conduction Properties of $CaTi_{1-x}M_xO_{3-\alpha}$ (M = Ga, Sc) at Elevated Temperatures. *Solid State Ionics*, *139*, 179-187.

[75] Huang, P., & Petric. A. (1996). Superior Oxide Ion Conductivity of lanthanum Gallate Doped with Strontium and Magnesium. *J. Electrochem. Soc.*, *143*, 1644-1648.

[76] Huang, K., Tichy, R. S., & Goodenough, J. B. (1998). Superior Perovskite Oxide Ion Conductor; Strontium- and Magnesium Doped $LaGaO_3$: I, Phase Relationships and Electrical Properties. *J. Am. Ceram. Soc.*, *81*, 2565-2575.

[77] Lybye, D., Poulsen, F. W., & Mogensen, M. (2000). Conductivity of A- and B-site Doped $LaAlO_3$, $LaGaO_3$, $LaScO_3$ and $LaInO_3$ Perovskites. *Solid State Ionics*, *128*, 91-103.

[78] Chen, F., & Liu, M. (1998). Study of Transition Metal Oxide Doped $LaGaO_3$ as Electrode Materials for LSGM-Based Solid Oxide Fuel Cells. *J. Solid State Electrochem.*, *3*, 7-14.

[79] Ishihara, T., Furutani, H., Honda, M., Yamada, T., Shibayama, T., Akbay, T., Sakai, N., Yokokawa, H., & Takita, Y. (1999). Improved Oxide Ion Conductivity in $La_{0.8}Sr_{0.2}Ga_{0.8}Mg_{0.2}O_3$ by Doping Co. *Chem. Mater.*, *11*, 2081-2088.

[80] Yamaji, K., Horita, T., Ishikawa, M., Sakai, N., & Yokokawa, H. (1998). Compatibility of $La_{0.9}Sr_{0.1}Ga_{0.8}Mg_{0.2}O_{2.85}$ as the Electrolyte for SOFCs. *Solid State Ionics*, *108*, 415-421.

[81] Yamaji, K., Horita, T., Ishikawa, M., Sakai, N., & Yokokawa, H. (1999). Compatibility of $La_{0.9}Sr_{0.1}Ga_{0.8}Mg_{0.2}O_{2.85}$ as the Electrolyte for SOFCs. *Solid State Ionics*, *121*, 217-224.

[82] Kuncewicz-Kupczyk, W., Kobertz, D., Miller, M., Singheiser, L., & Hilpert, K. (2001). Vaporization of Sr- and Mg-Doped Lanthanum Gallate and Implications for Solid Oxide Fuel Cells. *J. Electrochem. Soc.*, *148*, E276-E281.

[83] Weppner, W. (1977). Voltage Relaxation Measurements of the Electron and Hole Mobilities in Yttria-Doped Zirconia. *Electrochimica Acta*, *22*, 721-727.

[84] (a) Thangadurai, V., & Weppner, W. (2001). Synthesis and Electrical Properties of K- and Pr-Substituted $LaGaO_3$ and $LaInO_3$ Perovskites. *J. Electrochem. Soc.*, *148*, A1294-A1301. (b) Thangadurai, V., & Weppner, W. (2005). Studies on Electrical Properties of $La_{0.8}Sr_{0.2}Ga_{0.8}Mg_{0.2}O_{2.80}$ (LSGM) and LSGM–$SrSn_{1-x}Fe_xO_3$ (x = 0.8; 0.9) Composites and Their Chemical Reactivity. *Electrochimica Acta.*, *50*, 1871-1877.

[85] Kröger, F. A. (1974). The Chemistry of Imperfect Crystals Imperfection Chemistry of Crystalline Solids, Vol. 2., 2nd Edition, Amsterdam: North-Holland.

[86] Weppner, W., & Huggins, R. A. (1978). Electrochemical Methods for Determining Kinetic Properties of Solids. *Ann. Rev. Mater. Sci.*, *8*, 269-311.

[87] Weppner, W., & Huggins, R. A. (1977). Determination of the Kinetic Parameters of Mixed-Conducting Electrodes and Application to the System Li_3Sb. J. *Electrochem. Soc.*, *124*, 1569-1578.

[88] Roh, K. S., Ryu, K. H., & Yo, C. H. (1999). Non-stoichiometry and Physical Properties of the $SrSn_{1-x}Fe_xO_{3-y}$ System. *J. Solid State Chem. 142*, 288-293.

[89] Thangadurai, V., Schmid-Beurmann, P., & Weppner, W. (2002). $SrSn_{1-x}Fe_xO_{3-\delta}$ ($0 \leq x \leq 1$) Perovskites: a Novel Mixed Oxide Ion and Electronic Conductor. *Mater. Res. Bull, 37*, 599-604.

[90] Thangadurai, V., Schmid-Beurmann, P., & Weppner, W. (2003). Mixed Oxide Ion and Electronic Conductivity in Perovskite-Type $SrSnO_3$ by Fe Substitution. *Mater. Sci, Eng. B100*, 18-22.

[91] Thangadurai, V., Schmid-Beurmann, P., & Weppner, W. (2002). Phase Transitions in the $SrSnO_3$- $SrFeO_3$ Solid Solutions: X-ray Diffraction and Mössbauer Studies. *J. Solid State Chem. 174*, 392-402.

[92] Shannon, R. D. (1976). Revised Effective Ionic Radii and Systematic Studies of Interatomic Distances in Halides and Chalcogenides. *Acta Cryst. A32*, 751-767.

[93] Thangadurai, V., Huggins, R. A., & Weppner, W. (2002). Use of Simple ac Technique to Determine the Ionic and Electronic Conductivities in Pure and Fe-Substituted $SrSnO_3$ Perovskite. *J. Power Sources*, *108*, 64-69.

[94] Liu, J., & Weppner, W. (1999). Electronic Conductivity Measurement of Gadolinia Doped Ceria by DC-Partial-Polarization. *Ionics*, *5*, 115-121.

[95] Lübke, S., & Wiemhöfer, H.-D. (1999). Electronic Conductivity of Gd-Doped Ceria with Additional Pr-Doping. *Solid State Ionics*, *117*, 229-243.

[96] Hebb, M. H. (1952). Electrical Conductivity of Silver Sulfide. *J. Chem. Phys. 20*, 185-190.

[97] Wagner C. (1957). Galvanic Cells with Solid Electrolytes Involving Ionic and Electronic Conduction. In *Proc. Int. Comm. Electrochem. Thermodyn. and Kinetics* (CITCE) p.361-377. London: Butterworths.

[98] Thangadurai, V., & Weppner, W. (2004). $Ce_{0.8}Sm_{0.2}O_{1.9}$: Characterization of Electronic Charge Carriers and Application in Limiting Current Oxygen Sensors. *Electrochimica Acta, 49*, 1577-1585.

[99] Weppner, W. (1977). Electronic Transport Properties and Electrically Induced p-n Junction in ZrO_2 + 10 m/o Y_2O_3. *J. Solid State Chem.*, *20*, 305-314.

[100] Lübke, S., Wiemhöfer, H. –D. (1998). Electronic Conductivity of Gadolinia Doped Ceria. *Ber. Bunsenges. Phys. Chem. 102*, 642-649.

[101] Kim, J. H., Yoo, H. I. (2001). Partial Electronic Conductivity and Electrolytic Domain of $La_{0.9}Sr_{0.1}Ga_{0.8}Mg_{0.2}O_{3-\delta}$. *Solid State Ionics, 140*, 105-113.

[102] Schmidt, S., Berckemeyer, F., & Weppmer, W. (1999). Investigations of Electronic Minority Charge Carrier Conductivity in $La_{0.9}Sr_{0.1}Ga_{0.8}Mg_{0.2}O_{2.85}$. *Ionics, 6*, 139-144.

[103] Kopp, A., Näfe, H., & Weppner, W. (1992). Characterization of the Electronic Charge Carriers in TZP. *Solid State Ionics, 53-56*, 853-858.

[104] Xiong, Y., Yamaji, K., Horita, T., Sakai, N., & Yokokawa, H. (2002). Electronic Conductivity of 20 mol % $YO_{1.5}$ Doped CeO_2. *J. Electrochem. Soc. 149*, E450-E454.

[105] Liaw, B. Y., & Weppner, W. (1990). Low Temperature Limiting–Current Oxygen Sensors Using Tetragonal Zirconia as Solid Electrolytes. *Solid State Ionics, 40/41*, 428-432.

[106] Cotton, F. A. (1966). Advanced Inorganic Chemistry A Comprehensive Text, 2nd Edition, New York: Interscience Publishers.

[107] Hoffart, L., Heider, U., Jörissen, L., Huggins, R. A., & Witschel, W. (1995). Transport Properties of Materials with the Scheelite Structure and Their Modification by Doping, *Ionics*, *1*, 131-135.

INDEX

S

T

V

W

X

Z